PAULI LECTURES ON PHYSICS VOLUME 1

Electrodynamics

Wolfgang Pauli

Edited by Charles P. Enz

Translated by S. Margulies and H. R. Lewis

Foreword by Victor F. Weisskopf

DOVER PUBLICATIONS, INC.
Mineola, New York

Copyright

Bibliographical Note

This Dover edition, first published in 2000, is an unabridged republication of the work originally published in 1973 by The MIT Press, Cambridge, Massachusetts and London, England.

Library of Congress Cataloging-in-Publication Data

Pauli, Wolfgang, 1900–1958.
[Vorlesung über Elektrodynamik. English]
Electrodynamics / Wolfgang Pauli ; edited by Charles P. Enz ; translated by S. Margulies and H.R. Lewis ; foreword by Victor F. Weisskopf.
p. cm. — (Pauli lectures on physics ; v. 1)
Originally published: Cambridge, Mass. : MIT Press, 1973.
Includes bibliographical references and index.
ISBN 0-486-41457-4 (pbk.)
1. Electrodynamics. I. Enz, Charles P. (Charles Paul), 1925–
II. Title.

QC3 .P35 2000 vol. 1
[QC631]
530 s—dc21
[537.6]

00-031582

Manufactured in the United States of America
Dover Publications, Inc., 31 East 2nd Street, Mineola, N.Y. 11501

Pauli Lectures on Physics in Dover Editions

Volume 1. Electrodynamics (41457-4)
Volume 2. Optics and the Theory of Electrons (41458-2)
Volume 3. Thermodynamics and the Kinetic Theory of Gases (41461-2)
Volume 4. Statistical Mechanics (41460-4)
Volume 5. Wave Mechanics (41462-0)
Volume 6. Selected Topics in Field Quantization (41459-0)

Contents

Foreword

It is often said that scientific texts quickly become obsolete. Why are the Pauli lectures brought to the public today, when some of them were given as long as twenty years ago? The reason is simple: Pauli's way of presenting physics is never out of date. His famous article on the foundations of quantum mechanics appeared in 1933 in the German encyclopedia *Handbuch der Physik*. Twenty-five years later it reappeared practically unchanged in a new edition, whereas most other contributions to this encyclopedia had to be completely rewritten. The reason for this remarkable fact lies in Pauli's style, which is commensurate to the greatness of its subject in its clarity and impact. Style in scientific writing is a quality that today is on the point of vanishing. The pressure of fast publication is so great that people rush into print with hurriedly written papers and books that show little concern for careful formulation of ideas. Mathematical and instrumental techniques have become complicated and difficult; today most of the effort of writing and learning is devoted to the acquisition of these techniques instead of insight into important concepts. Essential ideas of physics are often lost in the dense forest of mathematical reasoning. This situation need not be so. Pauli's lectures show how physical ideas can be presented clearly and in good mathematical form, without being hidden in formalistic expertise.

Pauli was not an accomplished lecturer in the technical sense of the word. It was often difficult to follow his courses. But when the sequence of his thoughts and the structure of his logic become apparent, the attentive follower is left with a new and deeper knowledge of essential concepts and with a clearer insight into the splendid architecture of reason, which is theoretical physics. The value of the lecture notes is not diminished by the fact that they were written not by him but by some of his collaborators. They bear the mark of the master in their conceptual structure and their mathematical rigidity. Only here and there does one miss words and comments of the master. Neither does one notice the passing of time in his lectures, with the sole exception of the lectures on field quantization, in which some concepts are formulated in a way that may appear old-fashioned to some today. But even these lectures should be of use to modern students because of their compactness and their direct approach to the central problems.

May this volume serve as an example of how the concepts of theoretical physics were conceived and taught by one of the great men who created them.

Victor F. Weisskopf

Cambridge, Massachusetts

Preface

As stated in the introductory survey, Pauli chose to teach electrodynamics in the inductive way, in opposition to his teacher, Arnold Sommerfeld, who had preference for the axiomatic presentation of Maxwell's equations (see the preface in A. Sommerfeld, *Electrodynamics*, Academic Press, New York, 1952). This attitude is typical of Pauli and can also be found in the other lectures of this series (with the exception of field quantization). It expresses Pauli's vivid interest in the formation of scientific concepts and ideas and the logical structures built upon them—an eminently historic process (see, e.g., the introduction in W. Pauli, "Der Einfluss archetypischer Vorstellungen auf die Bildung naturwissenschaftlicher Theorien bei Kepler" in *Naturerklärung und Psyche*, Rascher Verlag, Zürich, 1952).

Because of this concern for the logical structure of the theory, the present lectures are still a rewarding source of learning for students of today. The notes prepared by A. Thellung, on which this English translation is based, reflect well the concise style of Pauli's lecturing. When these notes were published at ETH, Zürich, in 1949, Thellung was Pauli's Ph.D. student. He later became Pauli's assistant and is now Professor of theoretical physics at Zürich University. The care and precision of his notes, so typical of Thellung, made the work of the translators a comparatively easy job.

For Pauli the central problem of electrodynamics was the

field concept and the existence of an elementary charge which is expressible by the fine-structure constant $e^2/mc = 1/137$. This fundamental pure number had greatly fascinated Pauli, as can be seen from the list of references to his work assembled in the appendix. For Pauli the explanation of the number 137 was the test of a successful field theory, a test which no theory has passed up to now. This number 137 transcended into a magic symbol at Pauli's death. When I visited Pauli in the hospital, he asked me with concern whether I had noticed his room number: 137! It is in this room that he died a few days later.

Charles P. Enz

Geneva, 17 November 1971

Electrodynamics

PAULI LECTURES ON PHYSICS VOLUME 1

Survey of the Historical Development and the Current Problems of Electrodynamics

Electrodynamics is a relatively young branch of theoretical physics. As a domain of field physics it has become very important in recent times. Field physics is concerned with continuous functions in space and time (examples: elasticity theory, hydrodynamics, etc.), in contrast to corpuscular physics, which also plays a role in electrodynamics since electricity has an atomistic structure.

The concept of a field goes back to Faraday, who employed an intuitive approach without rigorous mathematical formulas, and who introduced the concept of lines of force. Maxwell brought the theory into a rigorous mathematical form. The existence of the electromagnetic waves predicted by this theory was later verified experimentally by Hertz. Since light is nothing but electromagnetic waves within a certain wavelength region, it became possible to consider optics as a branch of electrodynamics.

The mechanics of continuous media provided the model for field physics. Thus, by analogy, it was believed that an "aether" was required for the propagation of light, and an attempt was made to attribute the phenomena of the propagation of light to the mechanical properties of this aether. (Mechanistic concept of nature: all the laws of nature were to be explained on the basis of mechanics.) Maxwell, Boltzmann, and others worked out so-called aether engines. However, it was soon seen that all these theories became very artificial and complicated. Thus, little by little, this approach was abandoned and the concept of a physical field became firmly established. However, electromagnetic fields can be measured only with so-called test charges; hence they are linked with concepts of corpuscular physics.

It has become evident that the aether cannot possess any atomistic structure (as indeed there is in the case of elastic media), and that the concept of motion can be applied only to the test charge and not to the aether itself. This latter fact has led to the development of the theory of relativity.

It is, however, by no means true that field physics has triumphed over corpuscular physics, as can be seen from the fact that electricity is atomistic in nature. The carriers of electrical charge are negative electrons and positive protons. In addition, positive electrons (positrons), negative antiprotons, and other particles (mesons and hyperons) of both positive and negative charge can be artificially produced by suitable processes. All electrical charges are integral multiples of an elementary charge.

H. A. Lorentz and J. Larmor traced the differences in the electrical behavior of macroscopic bodies, characterized by so-called "material constants," back to Maxwell's equations in vacuum and to carriers of electric charge (electron theory). They thus achieved a great simplification of the foundations, which proved to be of special advantage for relativity theory. On the other hand, there is no explanation for the fact that only integral multiples of a certain charge occur. The existence of an elementary charge has, until now, in no way been made plausible. It is still an open problem in theoretical physics. The electron itself is a stranger in the Maxwell-Lorentz theory as well as in present-day quantum theory [A-1]. [1]

The field-particle description presents a conceptual problem: although a field can be described mathematically

[1] Comments [A-1]–[A-4] appear in the Appendix on pp. 155–156.

without the need for any test charges, it cannot be measured without them. On the other hand, the test charge itself gives rise to a field. However, it is impossible to measure an external field with a test charge and, at the same time, to determine the field due to this charge. A certain duality exists. Consequently, electrodynamics is of great significance for physical epistemology [A-1].

Electrodynamics can be presented in two ways:

1. Deductive: starting with the Maxwell equations and developing special cases.
2. Inductive: beginning with the basic laws obtained from experiment and eventually building up to the Maxwell equations. This way corresponds more closely to the historical development.

In these lectures, we will employ the second approach.

Chapter 1. Electrostatics and Magnetostatics

1. COULOMB'S LAW

The concept of conductors and insulators is assumed to be familiar from experimental physics. In these paragraphs we will consider only charged bodies whose linear dimensions are small as compared with their relative separation. Thus, we can assume that their electric charges are concentrated at points (*point charges*).

The forces between two such point charges are *central* and satisfy the principle of *action and reaction* (action=reaction). The forces can be either attractive or repulsive and their intensities are given by

$$K = \frac{|e_1 \cdot e_2|}{r^2},$$

where e_1 and e_2 are the electric charges and r is their separation. In vector notation (Fig. 1.1), the force on the

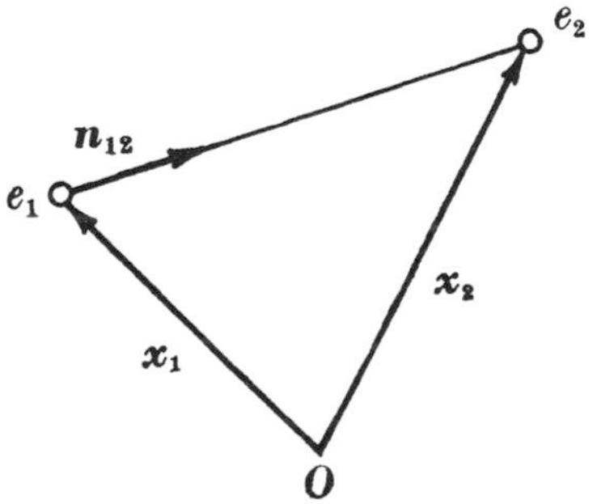

Figure 1.1

charge e_2 due to the charge e_1 is

$$\boldsymbol{K}_{12} = \frac{e_1 e_2}{r_{12}^2} \boldsymbol{n}_{12} = \frac{e_1 e_2}{r_{12}^3} (\boldsymbol{x}_2 - \boldsymbol{x}_1) = -e_1 e_2 \frac{\partial}{\partial \boldsymbol{x}_2} \frac{1}{r_{12}}, \qquad [1.1]$$

where $\boldsymbol{x}_1$ and $\boldsymbol{x}_2$ are the position vectors of e_1 and e_2, $r_{12} = |\boldsymbol{x}_2 - \boldsymbol{x}_1|$ is the separation between e_1 and e_2, $\boldsymbol{n}_{12} = (\boldsymbol{x}_2 - \boldsymbol{x}_1)/r_{12}$ is a unit vector in the direction from e_1 to e_2. This *empirically obtained* relation is called *Coulomb's law.*

One must add two other experimentally deduced facts:

1. *Superposition of forces.* Electric forces add according to the laws of vector addition (parallelogram rule):

$$\boldsymbol{K}_{1+2,3} = \boldsymbol{K}_{1,3} + \boldsymbol{K}_{2,3}.$$

2. *Conservation of charge.* Electric charges, which can be neither created nor destroyed, add algebraically.

Then Coulomb's law implicitly provides a sufficient definition of electrical charge since various quantities can be varied.

For example, one defines: (*a*) two charges to be equal when each produces the same effect upon a third charge; (*b*) a charge e_1 to be twice as large as charge e_2 if it produces the same effect upon a third charge e_3 as do two charges, each of magnitude e_2. The distance between these two charges e_2 must, of course, be small as compared with their distance to e_3. Electric charges can be either positive or negative. Charges of like sign repel, those of opposite sign attract.

One can also consider the forces exerted by a definite charge e_1 on various charges e_2, and vice versa. This dual aspect forms the basis for the *definition of the electric field*: For a given e_1, the force on a test charge e_2 is

$$\boldsymbol{K}_2 = e_2 \boldsymbol{E}. \qquad [1.2]$$

Here $\boldsymbol{E}$, the *electric field intensity*, is independent of e_2.

For the case of a field originating from a *single point*

charge e_Q at the point Q, the field at a point P is given by Coulomb's law:

$$\boldsymbol{E}_P = \frac{e_Q}{r_{QP}^2}\boldsymbol{n}_{QP} = \frac{e_Q}{r_{QP}^3}(\boldsymbol{x}_P - \boldsymbol{x}_Q) = -e_Q \frac{\partial}{\partial \boldsymbol{x}_P}\frac{1}{r_{QP}}, \qquad [1.3]$$

where Q, the source point, is fixed, and P, the field point, is arbitrary. For the case of *several point charges*, it follows from the principle of superposition that

$$\boldsymbol{E}_P = \sum_Q \frac{e_Q}{r_{QP}^2}\boldsymbol{n}_{QP} = \sum_Q \frac{e_Q}{r_{QP}^3}(\boldsymbol{x}_P - \boldsymbol{x}_Q) = -\frac{\partial}{\partial \boldsymbol{x}_P}\sum_Q \frac{e_Q}{r_{QP}}. \qquad [1.4]$$

In the *general case*, for any arbitrary charge distribution, the electric field intensity is also defined by the equation

$$\boldsymbol{K} = e\boldsymbol{E}. \qquad [1.5]$$

Here $\boldsymbol{K}$ is the force on the test body, e is the electric charge carried by the test body and depends only on the test body, and $\boldsymbol{E}$ is the electric field intensity which is independent of the test body and depends only on the field.

It is essential that we assume the existence of test charges that do not perturb the field or whose effects on the field are negligibly small. In macroscopic electrostatics this produces no difficulties since e, the magnitude of the test charge, can be made practically arbitrarily small. Furthermore, in the case that the sources of the field are fixed point or quasi-point charges, e can even be arbitrarily large. On the other hand, if the sources of the field are extended charged conducting surfaces, then a large test charge would produce a displacement of the charges, thereby changing the field (see Sec. 11). Contrary to macroscopic electrostatics, great difficulties arise in the atomic domain. Here, the reaction of the test charge on the field cannot be neglected since the test charge cannot be made arbitrarily small and, moreover, since the field sources are not at rest. This points to a definite difficulty in the field concept [A-1].

Up to this point we have described the field in the language of corpuscular physics. In the following sections we will consider the field defined by Coulomb's law as an independent concept and will investigate its properties.

2. THE FIELD OF POINT CHARGES

a. The field is conservative (conservation of energy)

From vector analysis we know that this property can be expressed by any of four fully equivalent expressions:

1. $$\oint \boldsymbol{E} \cdot \mathrm{d}\boldsymbol{s} = 0 \qquad [2.1]$$

for an arbitrary closed curve. The physical significance is

$$e \oint \boldsymbol{E} \cdot \mathrm{d}\boldsymbol{s} = \oint \boldsymbol{K} \cdot \mathrm{d}\boldsymbol{s} = 0\,,$$

that is, no net work is either gained or lost when a charged particle traverses a closed path in an electric field.

2. $\int_O^P \boldsymbol{E} \cdot \mathrm{d}\boldsymbol{s}$ is dependent only upon the position of the end points O and P (Fig. 2.1) and not on the path between

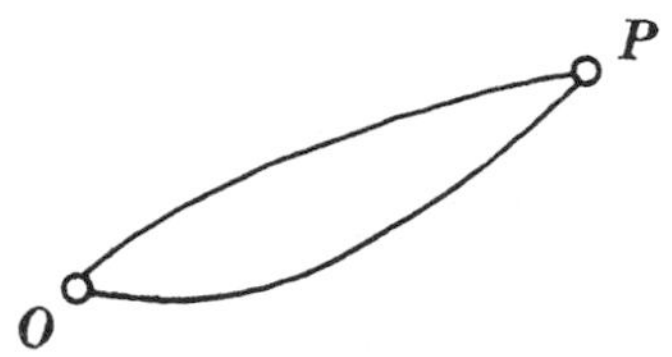

Figure 2.1

them. From this it can easily be shown that this integral must have the form

$$\int_O^P \boldsymbol{E} \cdot \mathrm{d}\boldsymbol{s} = -\varphi_P + \varphi_O\,. \qquad [2.2]$$

Proof: We define $\int_O^P \boldsymbol{E}\cdot d\boldsymbol{s} = F(O, P)$. Then, for an arbitrary point P',

$$F(O, P') + F(P', P) = F(O, P)$$

or

$$F(O, P') - F(P, P') = F(O, P)\,,$$

since obviously $F(P', P) = -F(P, P')$. We now keep P' fixed:

$$F(O, P') = \varphi_O \quad \text{and} \quad F(P, P') = \varphi_P\,.$$

Therefore,

$$\varphi_O - \varphi_P = F(O, P)\,,$$

which is the desired result.

The quantity φ_P is called the *electrostatic potential* at the point P. It is defined only to within an arbitrary additive constant. (The zero of potential can be chosen arbitrarily.)

3. $$\boldsymbol{E} = -\operatorname{grad}\varphi\,. \qquad [2.3]$$

In other notations,

$$\boldsymbol{E} = -\nabla\varphi; \qquad \boldsymbol{E} = -\frac{\partial\varphi}{\partial\boldsymbol{x}}; \qquad E_i = -\frac{\partial\varphi}{\partial x_i}, \qquad i = 1, 2, 3\,.$$

4. $$\operatorname{curl}\boldsymbol{E} = 0\,. \qquad [2.4]$$

Other notations are

$$\nabla\times\boldsymbol{E} = 0; \qquad \frac{\partial E_i}{\partial x_j} - \frac{\partial E_j}{\partial x_i} = 0\,, \qquad i, j \text{ cyclic}\,.$$

The equivalence of condition 4 with the three other forms of the energy law follows from *Stokes's theorem*:

$$\int_C \boldsymbol{E}\cdot d\boldsymbol{s} = \int_F \operatorname{curl}\boldsymbol{E}\cdot\boldsymbol{n}\,df\,. \qquad [2.5]$$

For this equation, in a right-handed coordinate system, the relation between the normal direction and the sense of rotation of C shown in Fig. 2.2 must be maintained.

From each of the above four statements the other three follow. They can be shown to be valid for any arbitrary central force. Thus, they are also valid for a finite sum

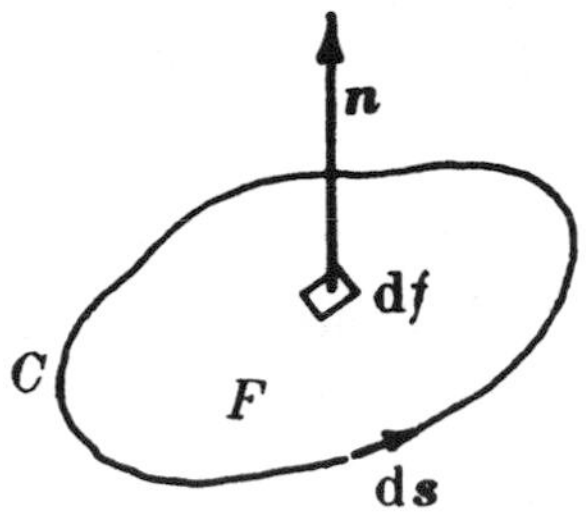

Figure 2.2

of central forces and, consequently, also for the field of point charges.

The potential due to a system of point charges is

$$\varphi_P = \sum_Q \frac{e_Q}{r_{QP}}. \qquad [2.6]$$

Here, we have chosen the arbitrary additive constant so that $\varphi(\infty) = 0$. The correctness of this expression for φ can be shown by forming the gradient:

$$\boldsymbol{E} = -\operatorname{grad}_P \varphi = \sum_Q \frac{e_Q}{r_{QP}^3} (\boldsymbol{x}_P - \boldsymbol{x}_Q) .$$

(The differentiation is with respect to the coordinates of P; the sources remain fixed.)

Since statements 1, 2, 3, and 4 are valid for arbitrary potential fields, they say far less about the electric field $\boldsymbol{E}$ than does Coulomb's law. They are equivalent to Coulomb's law only when taken together with the following law.

b. Electric field flux. Gauss's law

1. *A single point charge.* The electric field flux through a sphere with the source point at its center (Fig. 2.3) is

$$\oint_K E_n \, \mathrm{d}f = \int_K \frac{e}{r^2} r^2 \, \mathrm{d}\Omega = 4\pi e .$$

The flux is the same for all spheres regardless of their size. One can easily see by extension that for an *arbitrary closed surface*

$$\oint_F E_n \,\mathrm{d}f = -\oint_F \frac{\partial\varphi}{\partial n}\,\mathrm{d}f$$

$$= \begin{cases} 4\pi e & \text{if } Q \text{ lies inside the surface } F\,, \\ 0 & \text{if } Q \text{ lies outside the surface } F\,. \end{cases} \qquad [2.7]$$

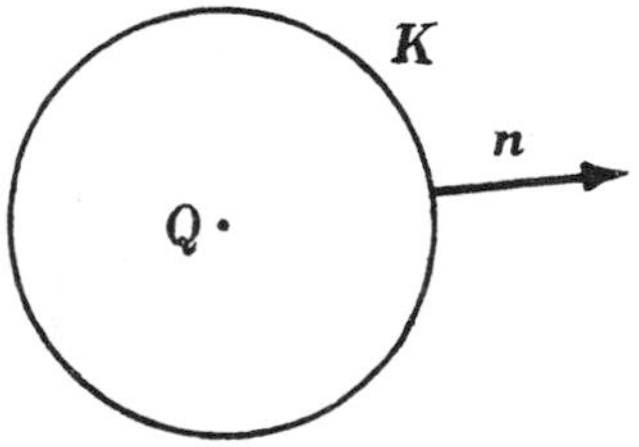

Figure 2.3

2. *Several point charges.* In this case,

$$\oint_F E_n \,\mathrm{d}f = -\oint_F \frac{\partial\varphi}{\partial n}\,\mathrm{d}f = 4\pi \sum_Q e_Q\,. \qquad [2.8]$$

The sum is to be taken over all the charges contained within the surface F.

Because of *Gauss's theorem*,

$$\oint_F E_n \,\mathrm{d}f = \oint_F \boldsymbol{E}\cdot\boldsymbol{n}\,\mathrm{d}f = \int_V \operatorname{div}\boldsymbol{E}\,\mathrm{d}V\,, \qquad [2.9]$$

in which F is a closed surface which bounds the volume V and $\boldsymbol{n}$ is the outward normal, it follows from Eqs. [2.7] and [2.8] that

$$\operatorname{div}_P \boldsymbol{E} = 0 \qquad \text{for} \qquad P \neq Q\,. \qquad [2.10]$$

Since $\boldsymbol{E} = -\operatorname{grad}\varphi$ and

$$\operatorname{div}\operatorname{grad}\varphi = \nabla^2\varphi = \Delta\varphi = \left(\frac{\partial^2}{\partial x_1^2} + \frac{\partial^2}{\partial x_2^2} + \frac{\partial^2}{\partial x_3^2}\right)\varphi\,,$$

φ satisfies *Laplace's equation*:

$$\nabla_P^2 \varphi = 0 \quad \text{for} \quad P \neq Q . \qquad [2.11]$$

It is immediately seen that $\varphi = e/r$ (for a single point charge) or $\varphi = \sum_Q e_Q / r_{QP}$ (for several point charges) are solutions of this equation, because

$$\frac{\partial r}{\partial x_i} = \frac{x_i}{r}; \qquad \frac{\partial}{\partial x_i}\left(\frac{1}{r}\right) = -\frac{x_i}{r^3}, \qquad \frac{\partial^2}{\partial x_i^2}\left(\frac{1}{r}\right) = \frac{3x_i^2 - r^2}{r^5},$$

$$\nabla^2\left(\frac{1}{r}\right) = \sum_i \frac{\partial^2}{\partial x_i^2}\left(\frac{1}{r}\right) = \frac{3r^2 - 3r^2}{r^5} = 0 ,$$

except at the origin. For a surface containing the origin,

$$-\oint_F \frac{\partial}{\partial n}\left(\frac{1}{r}\right) \mathrm{d}f = \oint_F \frac{1}{r^2}\left(\boldsymbol{n} \cdot \frac{\boldsymbol{x}}{r}\right) \mathrm{d}f = 4\pi .$$

Energy conservation,

$$\operatorname{curl} \boldsymbol{E} = 0 \quad \text{or} \quad \boldsymbol{E} = -\operatorname{grad} \varphi ,$$

and Gauss's law,

$$\oint_F E_n \,\mathrm{d}f = 4\pi \sum_Q e_Q \quad \text{or} \quad -\oint \frac{\partial \varphi}{\partial n} \mathrm{d}f = 4\pi \sum_Q e_Q ,$$

(Q within F) are, together, completely equivalent to Coulomb's law for point charges.

Mathematical remarks on Stokes's and Gauss's theorems

In the case of Stokes's theorem we had a definite relation between the direction of the normal $\boldsymbol{n}$ and the sense of traversal of the curve C. This convention is necessary only if curl $\boldsymbol{E}$ and the surface element $\mathrm{d}\boldsymbol{f}$ are considered as vectors. It is, however, more consistent to consider them as antisymmetric tensors.[1] This can be seen by considering an

[1] For the concept of a tensor, see Section 18.

inversion of the coordinate system. However, to any antisymmetric tensor in a three-dimensional space,

$$C_{ik} = A_i B_k - A_k B_i = -C_{ki}$$

(the A_i and B_i are to transform as vector components), a vector C_l can be assigned with the help of the tensor ε_{ikl}, which is antisymmetric in all three indices:

$$\varepsilon_{ikl} = \begin{cases} \pm 1 & \text{if } ikl \text{ is an } \begin{pmatrix}\text{even}\\ \text{odd}\end{pmatrix} \text{ permutation of } 1,2,3, \\ 0 & \text{if at least two of the three indices are equal.} \end{cases}$$

That is, one sets

$$C_i = \sum_{k<l} \varepsilon_{ikl} C_{kl} .$$

Then, the vector components C_1, C_2, and C_3 are equal to the tensor components C_{23}, C_{31}, and C_{12}, respectively, in all coordinate systems which can be derived by proper rotations (i.e., where the determinant of the rotation matrix is $+1$) of the original system. If improper rotations (with determinant -1) are considered, then the signs change. For example,

$$x_i' = -x_i \qquad (i = 1, 2, 3),$$

$$C_{kl}' = C_{kl}, \quad \text{but} \quad C_i' = -C_i .$$

From the point of view of inversions, both the curl and the surface element can be considered as antisymmetric tensors (vector products). Thus, one can write

$$\oint \boldsymbol{E}\cdot d\boldsymbol{s} = \sum_{i<k} \int \operatorname{curl}_{ik} \boldsymbol{E}\, df_{ik} . \qquad [2.12]$$

Hence, an ordering between the normal direction and the sense of rotation is not necessary. Indeed, the concept of a normal never appears. In this form Stokes's theorem is valid in an arbitrary n-dimensional space. This very general theorem is *purely topological*; that is, it is independent of any metric. It is only necessary that the E_i transform contragrediently to the dx^i. That is,

$$\boldsymbol{E}\cdot d\boldsymbol{s} = \sum_i E_i\, dx^i = \text{invariant}.$$

A closed surface can always be thought of as being composed of two separate surfaces with a common boundary. If Eq. [2.12] is applied to both of these surfaces, then the line integrals along the bounding curves cancel and it follows, for a closed surface, that

$$\sum_{i<k} \oint \text{curl}_{ik}\, \boldsymbol{E}\, df_{ik} = 0\,.$$

The proof of Stokes's theorem proceeds most easily by first proving it for a rectangle and then showing that the theorem is invariant under arbitrary coordinate transformations. Thus, it must be true for all surfaces that can be transformed into a rectangle. For the case of multiply-connected regions, a separation into singly-connected regions, for each of which the theorem then applies, is made by appropriate cuts. Since the line integrals cancel in pairs along the cuts, the theorem is thus also valid for multiply-connected regions.

Starting with a vector A_i and an antisymmetric tensor $C_{ik} = -C_{ki}$, one can construct the tensor

$$A_i C_{kl} + A_k C_{li} + A_l C_{ik} = D_{ikl}\,.$$

Since D_{ikl} is antisymmetric in all three indices it can always be written in the form

$$D_{ikl} = \varepsilon_{ikl} D\,.$$

In Gauss's theorem, the concept of the normal component

can now be eliminated by writing

$$\oint_F (E_i\,\mathrm{d}f_{kl} + E_k\,\mathrm{d}f_{li} + E_l\,\mathrm{d}f_{ik})$$

$$= \int_V \operatorname{div} \boldsymbol{E} (\mathrm{d}V)_{ikl} = \int_V \operatorname{div} \boldsymbol{E}\,|\mathrm{d}V|\,\varepsilon_{ikl}\,. \qquad [2.13]$$

It is seen that Gauss's theorem is also of topological character. The left-hand side transforms as a determinant, as does the right-hand side, since div$\boldsymbol{E}$ is a scalar. As is true for Stokes's theorem, Gauss's theorem is tied neither to a 3-dimensional space nor to the concept of a metric.

As in the case of Stokes's theorem, Gauss's theorem is most easily proved by showing it to be true for a cube, then showing that it is invariant under arbitrary coordinate transformations, and so forth.

3. VOLUME AND SURFACE CHARGE DISTRIBUTIONS

It is often more convenient to think of charge distributions as being continuous instead of consisting of a series of point charges. As long as one remains in the domain of macroscopic physics, the actual atomic structure of electricity can be neglected.

a. Volume charges

For the charges contained within a volume V we imagine the following limiting process:

$$\begin{cases} \text{the number of charges} \to \infty, \\ \text{the magnitude of each individual charge} \to 0, \end{cases}$$

to take place in such a manner that the total charge remains constant. Thus,

$$\sum_Q e_Q \to e = \int_V \varrho\,\mathrm{d}V \qquad \text{(total charge in } V\text{)}.$$

The continuous function thus defined, $\varrho = \lim_{V\to 0}(e/V)$, is called the ***volume charge density.*** The potential φ_P of Eq. [2.6] and the field $\boldsymbol{E}_P$ of Eq. [1.4] then become

$$\varphi_P = \int \frac{\varrho_Q}{r_{QP}} \, \mathrm{d}V_Q \,, \qquad [3.1]$$

$$\boldsymbol{E}_P = -\operatorname{grad}_P \varphi_P = \int \frac{\varrho_Q}{r_{QP}^3} (\boldsymbol{x}_P - \boldsymbol{x}_Q) \, \mathrm{d}V_Q \,. \qquad [3.2]$$

In both of these integrals there are no singularities as the point P approaches Q, contrary to what occurs in the summation of point charges. Stated very roughly, the integrands ϱ_Q/r_{QP} and $\varrho_Q(\boldsymbol{x}_P - \boldsymbol{x}_Q)/r_{QP}^3$ become infinite more slowly than the volume element $\mathrm{d}V$ approaches zero. This can easily be shown to be rigorously true by an exact consideration of the limit. Thus, for continuous volume distributions of charge, ***E** and φ are regular inside the source region.*

For the case of volume charges, Gauss's law becomes

$$-\oint_F \frac{\partial \varphi}{\partial n} \, \mathrm{d}f = \oint_F E_n \, \mathrm{d}f = 4\pi \int_V \varrho \, \mathrm{d}V \,. \qquad [3.3]$$

Because of the regularity of the integrand, this can, with the aid of Gauss's theorem (Eq. [2.9]), be transformed into

$$-\int_V \nabla^2 \varphi \, \mathrm{d}V = \int_V \operatorname{div} \boldsymbol{E} \, \mathrm{d}V = 4\pi \int_V \varrho \, \mathrm{d}V \,.$$

Since this is true for any arbitrary volume V, and in particular for arbitrary small volumes, then

$$\operatorname{div} \boldsymbol{E} = 4\pi\varrho \qquad [3.4]$$

or

$$-\nabla^2 \varphi = 4\pi\varrho \,. \qquad [3.5]$$

Equation [3.5], which is a generalization of Laplace's equation (Eq. [2.11]), is called ***Poisson's equation.***

b. *Surface charges*

One can imagine a process exactly analogous to that employed for volume charges applied to a two-dimensional surface. Thus, one arrives at the concept of a ***surface charge density*** ω. In analogy,

$$e=\int_F \omega\, \mathrm{d}f \quad \text{(total charge on the surface)},$$

$$\varphi_P=\int \frac{\omega_Q}{r_{QP}}\mathrm{d}f_Q, \tag{3.6}$$

$$\boldsymbol{E}_P=-\operatorname{grad}_P\varphi=\int \frac{\omega_Q}{r_{QP}^3}(\boldsymbol{x}_P-\boldsymbol{x}_Q)\,\mathrm{d}f_Q. \tag{3.7}$$

As the field point P passes into the surface, φ remains finite (integrand $\sim 1/r$, surface element $\sim r^2$). *Thus φ is continuous at a charged surface.*

On the other hand, this is not the case for the component of the electric field normal to the surface. This is seen by applying Gauss's law of Eq. [2.9] to an imaginary

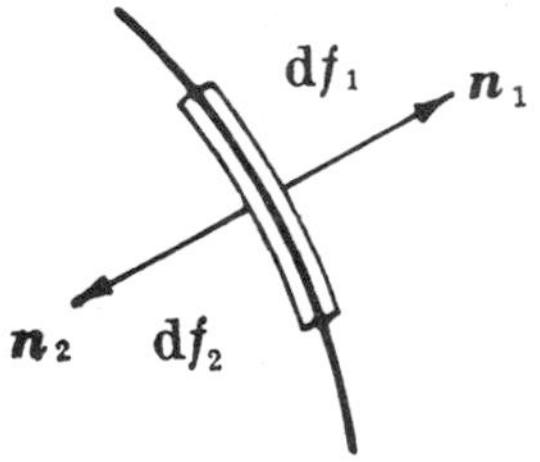

Figure 3.1

cylindrical surface containing a section of the charged surface, as shown in Fig. 3.1:

$$\oint E_n\,\mathrm{d}f=4\pi\int\omega\,\mathrm{d}f. \tag{3.8}$$

In the limit, as the height of the cylinder goes to zero while the base remains constant,

$$\int (E_{n_1}+E_{n_2})\,\mathrm{d}f=4\pi\int\omega\,\mathrm{d}f.$$

Since the base area is arbitrary,

$$E_{n_1} + E_{n_2} = 4\pi\omega \,. \qquad [3.9]$$

Alternately, with a different definition of the normal (Fig. 3.2),

$$(E_n)_1 - (E_n)_2 = 4\pi\omega \,,$$

$$\left(\frac{\partial\varphi}{\partial n}\right)_1 - \left(\frac{\partial\varphi}{\partial n}\right)_2 = -\,4\pi\omega \,.$$

Thus, the ***normal component of* E *is discontinuous*** at a charged surface. On the other hand, the ***tangential component remains continuous.*** ***Proof***: φ is the same on either side of the surface. Thus, differentiation parallel to the surface in the same direction must yield the same result, both inside and out. The proof can also be made with Stokes's theorem.

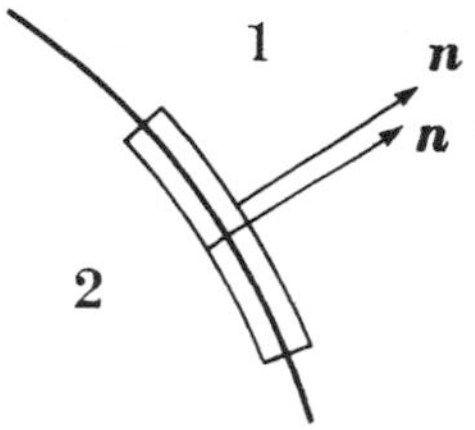

Figure 3.2

Boundary conditions for the field at conductors: Electrical conductors are substances with freely movable electrons. Electric force components inside a conductor produce charge displacements—a current flows. The conditions for no current flow or, equivalently, for the static character of the field, are

1. no field may be present inside the conductor:

$$\mathbf{E} = 0 \,;$$

2. at the surface of the conductor the tangential component

of the field vanishes:

$$E_{\parallel} = 0 \, .$$

(Since the tangential component is continuous, it must be zero outside the conductor just as it is inside.) On the contrary, the normal component outside the conductor is, in general, not zero since surface charges may be present. Thus, $\boldsymbol{E}$ inside is zero while externally the field consists of just the normal component

$$E_n = -\frac{\partial \varphi}{\partial n} = 4\pi\omega \, .$$

The gradient of the potential is perpendicular to the surface. Thus, $\varphi =$ constant along this surface. That is, the surface is an *equipotential surface.*

4. ENERGY OF THE ELECTROSTATIC FIELD

a. Point charges

The potential energy of a system of point charges is equal to the work required to bring the charges from infinity to their actual configuration. For two charges e_1 and e_2 at a distance r_{12} the work is

$$E_{\text{pot}} = \frac{e_1 e_2}{r_{12}} \, .$$

For an *arbitrary number of point charges*, by calculating the potential energy for each pair and summing over all pairs,

$$E_{\text{pot}} = \sum_{i<k} \frac{e_i e_k}{r_{ik}} \, . \qquad [4.1]$$

One can also write

$$E_{\text{pot}} = \frac{1}{2} \sum_{i,k}{}' \frac{e_i e_k}{r_{ik}} \, . \qquad [4.2]$$

The prime on the summation symbol is to indicate that

terms with $i = k$ are not included. From this symmetrical expression it can be seen that the energy thus defined is independent of the order in which the point charges are brought from infinity.

There is still a third form of E_{pot} in which the concept of potential is used. The potential at the position of the ith charge due to all charges except the ith is

$$\varphi_i' = \sum_{\substack{k \\ (k \neq i)}}' \frac{e_k}{r_{ik}} . \qquad [4.3]$$

Consequently,

$$E_{pot} = \frac{1}{2} \sum_i e_i \varphi_i' . \qquad [4.4]$$

b. Volume charges

In this case

$$E_{pot} = \frac{1}{2} \iint \frac{\varrho_P \varrho_{P'}}{r_{PP'}} \, dV_P \, dV_P . \qquad [4.5]$$

Here, the complication of $i \neq k$, which exists for point charges, does not appear, since in the expression $\int (\varrho_{P'}/r_{PP'}) dV_{P'}$ it makes no difference whether the volume element containing P is included or not.

In the notation using the potential,

$$\varphi_P = \int \frac{\varrho_{P'}}{r_{PP'}} \, dV_{P'} , \qquad [4.6]$$

$$E_{pot} = \frac{1}{2} \int \varrho_P \varphi_P \, dV_P . \qquad [4.7]$$

c. Volume charges and surface charges

With the potential

$$\varphi_P = \int \frac{\varrho_{P'}}{r_{PP'}} \, dV_{P'} + \int \frac{\omega_{P'}}{r_{PP'}} \, df_{P'} \quad \text{(Eqs. [4.6] plus [3.6])},$$

there results

$$E_{pot} = \frac{1}{2} \int \varrho_P \varphi_P \, dV_P + \frac{1}{2} \int \omega_P \varphi_P \, df_P . \qquad [4.8]$$

The first integral extends over all regions which contain volume charge distributions, the second over all charged surfaces. Without an explicit introduction of the potential concept,

$$E_{\text{pot}} = \frac{1}{2}\iint \frac{\varrho_P \varrho_{P'}}{r_{PP'}}\, \mathrm{d}V_P\, \mathrm{d}V_{P'} + \iint \frac{\varrho_P \omega_{P'}}{r_{PP'}}\, \mathrm{d}V_P\, \mathrm{d}f_{P'} + \frac{1}{2}\iint \frac{\omega_P \omega_{P'}}{r_{PP'}}\, \mathrm{d}f_P\, \mathrm{d}f_{P'}\,. \qquad [4.9]$$

As a consequence of the laws of electrostatics, these expressions for energy can be transformed into integrals over all space.

First, for *volume charges only*, using Eq. [3.4],

$$E_{\text{pot}} = \frac{1}{2}\int \varrho\varphi\, \mathrm{d}V = \frac{1}{8\pi}\int \varphi \operatorname{div} \boldsymbol{E}\, \mathrm{d}V\,.$$

Now, with the familiar identity from vector analysis,

$$\operatorname{div}(\varphi\boldsymbol{E}) = \varphi \operatorname{div}\boldsymbol{E} + \boldsymbol{E}\cdot\operatorname{grad}\varphi\,, \qquad [4.10]$$

which is valid for an arbitrary vector $\boldsymbol{E}$ and an arbitrary scalar function φ, Gauss's theorem yields

$$\oint \varphi E_n\, \mathrm{d}f = \int \varphi \operatorname{div}\boldsymbol{E}\, \mathrm{d}V + \int \boldsymbol{E}\cdot\operatorname{grad}\varphi\, \mathrm{d}V\,,$$

provided that $\operatorname{div}\boldsymbol{E}$ exists over the whole region of integration. (Thus, $\boldsymbol{E}$ must be continuous.) Applied to our situation, this results in

$$E_{\text{pot}} = \frac{1}{8\pi}\left\{-\int \boldsymbol{E}\cdot\operatorname{grad}\varphi\, \mathrm{d}V + \oint_K \varphi E_n\, \mathrm{d}f\right\}.$$

As the region of integration we imagine a sphere K whose radius r approaches infinity in the limit. We now assume that $\varrho \neq 0$ only within a finite sphere, or else that ϱ vanishes sufficiently fast at infinity so that the second integral over the sphere K vanishes as $K \to \infty$. (If $\varrho \neq 0$ only within a finite region, then $\varphi \to 0$ at least as fast as $1/r$ and $|\boldsymbol{E}| \to 0$ at least as $1/r^2$, while the surface increases only as r^2. Thus, the whole integral approaches zero at least as $1/r$.) Because $\boldsymbol{E} = -\operatorname{grad}\varphi$, one thus obtains

$$E_{\text{pot}} = \frac{1}{8\pi}\int \boldsymbol{E}^2 \,\mathrm{d}V\,, \qquad [4.11]$$

in which *the energy is expressed in terms of only the field intensity at all points in space.* The quantity $\boldsymbol{E}^2/8\pi$ is called the *energy density* of the field.

If *surface charges* are also present, the corresponding surfaces F must be excluded from the region of integration since

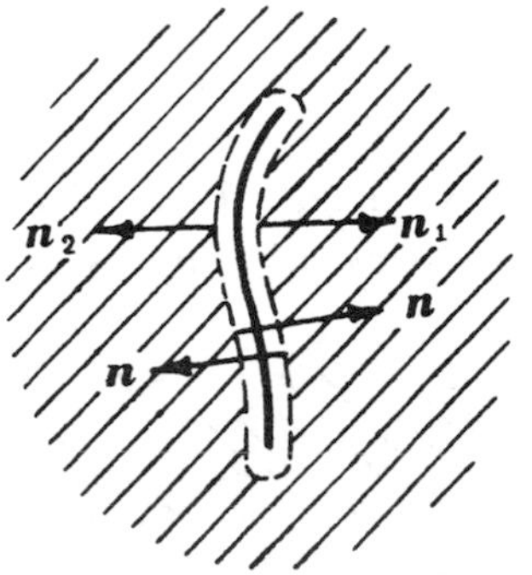

Figure 4.1

there $\boldsymbol{E}$ is not continuous and Gauss's theorem is not applicable. This gives us two additional terms which, however, cancel:

$$E_{\text{pot}} = \frac{1}{2}\int \varrho\varphi\,\mathrm{d}V + \frac{1}{2}\int \omega\varphi\,\mathrm{d}f$$

$$= \frac{1}{8\pi}\left\{-\int (\boldsymbol{E}\cdot\operatorname{grad}\varphi)\,\mathrm{d}V + \oint_K \varphi E_n\,\mathrm{d}f + \oint_F \varphi E_n\,\mathrm{d}f\right\} + \frac{1}{2}\int \omega\varphi\,\mathrm{d}f\,.$$

Now,

$$\frac{1}{8\pi}\oint_F \varphi E_n\,\mathrm{d}f = -\frac{1}{8\pi}\int \varphi(E_{n_1} + E_{n_2})\,\mathrm{d}f = -\frac{1}{2}\int \omega\varphi\,\mathrm{d}f\,,$$

since the normal $\boldsymbol{n}$, which must point away from the region of integration, is oppositely directed to $\boldsymbol{n}_1$ and $\boldsymbol{n}_2$ (see

Fig. 4.1). If we again assume that the integral over the sphere vanishes as $K \to \infty$, then

$$E_{\text{pot}} = \frac{1}{8\pi} \int \boldsymbol{E}^2 \, \mathrm{d}V \qquad [4.11]$$

is again obtained. This formula is thus valid *for both volume and surface charges.*

On the other hand, this integral diverges for *point charges.* In this case, the field must be decomposed into components arising from the various point charges. The potential and the field at the point P produced by the point charge e_k are

$$\varphi_k(P) = \frac{e_k}{r_P}, \qquad \boldsymbol{E}_k = -\operatorname{grad} \varphi_k .$$

We now consider the integral $\int \boldsymbol{E}_i \cdot \boldsymbol{E}_k \, \mathrm{d}V$ for $i \neq k$. If we exclude the singularities by enclosing the point charges

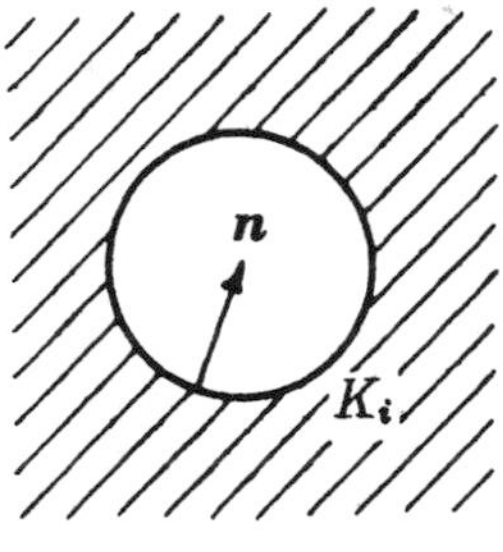

Figure 4.2

within small spheres K_i and K_k, then, remembering that div $\boldsymbol{E} = 0$ over the region of integration, Gauss's theorem yields

$$\begin{aligned}\int \boldsymbol{E}_i \cdot \boldsymbol{E}_k \, \mathrm{d}V &= -\int \boldsymbol{E}_i \cdot \operatorname{grad} \varphi_k \, \mathrm{d}V \\ &= -\int \operatorname{div} (\varphi_k \boldsymbol{E}_i) \, \mathrm{d}V + \int \varphi_k \operatorname{div} \boldsymbol{E}_i \, \mathrm{d}V \\ &= -\oint_K E_{in} \varphi_k \, \mathrm{d}f - \oint_{K_i} E_{in} \varphi_k \, \mathrm{d}f - \oint_{K_k} E_{in} \varphi_k \, \mathrm{d}f .\end{aligned}$$

The first term approaches zero as the large sphere becomes infinite. The third term approaches zero as we let K_k go to zero. Only the second term contributes. For $K_i \to 0$,

$$\varphi_k \to \frac{e_k}{r_{ik}},$$

$$-E_{in} = \frac{e_i}{r_{ik}^2},$$

$$-\oint_{K_i} E_{in}\,\mathrm{d}f = 4\pi e_i .$$

Thus,

$$-\oint_{K_i} E_{in}\varphi_k\,\mathrm{d}f \to 4\pi\frac{e_i e_k}{r_{ik}} .$$

For *two point charges*, therefore,

$$E_{\text{pot}} = \frac{e_i e_k}{r_{ik}} = \frac{1}{4\pi}\int \mathbf{E}_i\cdot\mathbf{E}_k\,\mathrm{d}V . \qquad [4.12]$$

Here, the integral extends over all space.

For an *arbitrary number of point charges*

$$E_{\text{pot}} = \sum_{i<k}\frac{e_i e_k}{r_{ik}} = \frac{1}{4\pi}\sum_{i<k}\int \mathbf{E}_i\cdot\mathbf{E}_k\,\mathrm{d}V = \frac{1}{8\pi}\sum_{\substack{i,k\\(i\neq k)}}\int \mathbf{E}_i\cdot\mathbf{E}_k\,\mathrm{d}V . \qquad [4.13]$$

One can also write

$$E_{\text{pot}} = \frac{1}{8\pi}\int\Big(\mathbf{E}^2 - \sum_i \mathbf{E}_i^2\Big)\,\mathrm{d}V . \qquad [4.14]$$

(In this case, however, the integral cannot be decomposed into two individual integrals since these would diverge.)

5. EXAMPLE: CHARGE DISTRIBUTIONS WITH SPHERICAL SYMMETRY

We consider a *volume distribution of charge* $\varrho(r)$ which depends only upon the distance r from a fixed point O. Thus, the potential $\varphi(r)$ is a function of r only.

In this case it is simplest to start from Poisson's equation

$$-\nabla^2\varphi = 4\pi\varrho\,. \qquad [5.1]$$

In polar coordinates the operator ∇^2, when applied to a function φ dependent only upon r, has the form

$$\nabla^2\varphi \equiv \frac{1}{r^2}\frac{\mathrm{d}}{\mathrm{d}r}\left(r^2\frac{\mathrm{d}\varphi}{\mathrm{d}r}\right) \equiv \frac{1}{r}\frac{\mathrm{d}^2}{\mathrm{d}r^2}(r\varphi) \equiv \frac{\mathrm{d}^2\varphi}{\mathrm{d}r^2} + \frac{2}{r}\frac{\mathrm{d}\varphi}{\mathrm{d}r}\,. \qquad [5.2]$$

Thus, we have

$$-\frac{1}{r^2}\frac{\mathrm{d}}{\mathrm{d}r}\left(r^2\frac{\mathrm{d}\varphi}{\mathrm{d}r}\right) = 4\pi\varrho\,,$$

$$-r^2\frac{\mathrm{d}\varphi}{\mathrm{d}r} = 4\pi\int_0^r \varrho r^2\,\mathrm{d}r \equiv e(r)\,,$$

where $e(r)$ is the charge contained within a sphere of radius r. Hence,

$$-\frac{\mathrm{d}\varphi}{\mathrm{d}r} = E_r = \frac{e(r)}{r^2}\,. \qquad [5.3]$$

The field at a point located a distance r from O, the center of the charge distribution, depends only upon the charge contained within a sphere of radius r about O. The charge outside does not affect $\boldsymbol{E}$ as long as it is distributed with spherical symmetry. The field is the same as it would be if all of the charge $e(r)$ were concentrated at the center O. The potential is given by

$$\varphi(r) = -\int_0^r \frac{e(r)}{r^2}\,\mathrm{d}r = \int_0^r e(r)\,\mathrm{d}\left(\frac{1}{r}\right)$$

$$= e(r)\frac{1}{r}\Big|_0^r - \int_0^r \frac{1}{r}\,e'(r)\,\mathrm{d}r\,.$$

Because $e'(r) = (\mathrm{d}/\mathrm{d}r)[e(r)] = 4\pi r^2\varrho$, and since, for $r \to 0$,

$e(r) \to 0$ as r^3, we have

$$\varphi(r) = \frac{e(r)}{r} - 4\pi \int_0^r \varrho r \, \mathrm{d}r + \text{constant}. \qquad [5.4]$$

a. Uniformly charged spherical volume

$$\varrho = \begin{cases} \varrho_0 = \text{constant} & \text{for } r < a \\ 0 & \text{for } r \geqslant a, \end{cases}$$

$$e(r) = \begin{cases} \frac{4}{3}\pi r^3 \varrho_0 = e\frac{r^3}{a^3} & r < a \\ \frac{4}{3}\pi a^3 \varrho_0 \equiv e & r \geqslant a, \end{cases}$$

$$-\frac{\mathrm{d}\varphi}{\mathrm{d}r} = E_r = \begin{cases} e\frac{r}{a^3} & r < a \\ \frac{e}{r^2} & r \geqslant a, \end{cases}$$

$$\varphi = \begin{cases} \frac{e}{2a^3}(3a^2 - r^2) & r < a \\ \frac{e}{r} & r \geqslant a. \end{cases}$$

(The constants have been chosen so that $\varphi(\infty) = 0$ and φ is continuous at $r = a$.) Calculation of the electrostatic energy yields

$$E_{\text{pot}} = \frac{1}{2}\int \varrho\varphi \, \mathrm{d}V = \frac{1}{2}\varrho_0 \frac{1}{2}\frac{e}{a^3}\int_0^a (3a^2 - r^2)4\pi r^2 \, \mathrm{d}r$$

$$= \frac{1}{2}\cdot\frac{3e}{a^3}\cdot\frac{1}{2}\cdot\frac{e}{a^3}\left[3a^2\frac{a^3}{3} - \frac{a^5}{5}\right] = \frac{3}{5}\cdot\frac{e^2}{a}.$$

The same result is obtained by using $\frac{1}{8\pi}\int E_r^2 \, \mathrm{d}V$.

b. Uniformly charged spherical surface

Let

$$\omega = \frac{e}{4\pi a^2},$$

where e is the total charge and a is the radius of the sphere. As a consequence of Gauss's law,

$$r^2 \cdot E_r = \text{constant}$$

must be valid both inside and outside the sphere. If we denote the region inside the sphere by 1 and the region outside by 2, then

$$r < a: \qquad E_{r1} = \frac{c_1}{r^2},$$

but $c_1 = 0$, since there can be no discontinuity at the center. Hence

$$E_{r1} = 0 .$$

$$r > a: \qquad E_{r2} = \frac{c_2}{r^2};$$

but

$$E_{r2} - E_{r1} = 4\pi\omega = \frac{e}{a^2} = (E_{r2})_{r=a} .$$

Hence, $c_2 = e$,

$$r > a: \qquad E_r = \frac{e}{r^2}, \qquad \varphi = \frac{e}{r};$$

$$r < a: \qquad E_r = 0, \qquad \varphi = \frac{e}{a}.$$

Again, we see that the surface charge acts externally as though it were concentrated at the center of the sphere. Within the sphere it produces no effect. The energy is

$$E_{\text{pot}} = \frac{1}{8\pi}\int E^2 \,\mathrm{d}V = \frac{e^2}{8\pi}\int_a^\infty \frac{4\pi r^2}{r^4}\,\mathrm{d}r = \frac{e^2}{2a}.$$

For the limiting case of a point charge, $E_{\text{pot}} \to \infty$.

6. PROOF OF THE EQUIVALENCE OF THE ELECTROSTATIC FIELD EQUATIONS WITH COULOMB'S LAW

We have derived the field equations [2.4], [2.3], or [3.4] [3.5],

$$\operatorname{curl} \boldsymbol{E} = 0, \qquad \operatorname{div} \boldsymbol{E} = 4\pi\varrho,$$

or

$$\boldsymbol{E} = -\operatorname{grad}\varphi, \qquad \nabla^2\varphi = -4\pi\varrho,$$

from Coulomb's law. We now wish to show that, conversely, Coulomb's law follows from these equations. In order to demonstrate this we must, in addition, require that $\varphi \to 0$ at least as fast as $1/r$ as $r \to \infty$.

For simplicity we will perform the proof only for *volume charges*. For this, we will need *Green's theorem*. If $\boldsymbol{A} = \varphi \operatorname{grad}\psi$ is substituted into Gauss's theorem,

$$\oint_F A_n \, \mathrm{d}f = \oint_F \boldsymbol{A} \cdot \boldsymbol{n} \, \mathrm{d}f = \int_V \operatorname{div} \boldsymbol{A} \, \mathrm{d}V,$$

then, because

$$\operatorname{div}(\varphi \operatorname{grad}\psi) = \varphi \nabla^2\psi + \operatorname{grad}\varphi \cdot \operatorname{grad}\psi, \qquad [6.1]$$

we obtain Green's first identity:

$$\oint \varphi \frac{\partial\psi}{\partial n} \, \mathrm{d}f = \int \varphi \nabla^2\psi \, \mathrm{d}V + \int \operatorname{grad}\varphi \cdot \operatorname{grad}\psi \, \mathrm{d}V. \qquad [6.2]$$

Interchanging φ and ψ in [6.1] we obtain, upon subtraction, Green's second identity:

$$\operatorname{div}(\varphi \operatorname{grad}\psi - \psi \operatorname{grad}\varphi) = \varphi \nabla^2\psi - \psi \nabla^2\varphi,$$

$$\oint \left(\varphi \frac{\partial\psi}{\partial n} - \psi \frac{\partial\varphi}{\partial n}\right) \mathrm{d}f = \int (\varphi \nabla^2\psi - \psi \nabla^2\varphi) \, \mathrm{d}V. \qquad [6.3]$$

These results are valid for any arbitrary volume in which φ and ψ are regular.

Green's second identity can be applied to the integration of Poisson's equation by considering φ as the sought potential function and letting $\psi_{P'} = 1/r_{PP'}$. In order that the integrands be regular, we limit our integration to the region

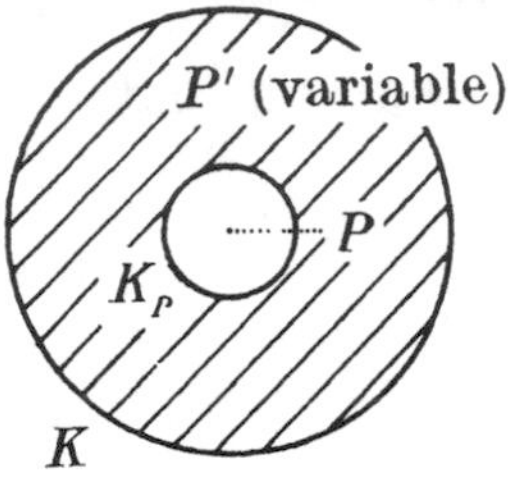

Figure 6.1

between a small sphere K_P about the fixed point P and a large sphere K (see Fig. 6.1). Because

$$\nabla^2 \psi = 0 \quad \text{for} \quad P \neq P',$$

we have

$$\oint_K \left(\varphi \frac{\partial(1/r)}{\partial n} - \frac{1}{r}\frac{\partial \varphi}{\partial n} \right) \mathrm{d}f_{P'} + \oint_{KP} \left(\varphi \frac{\partial(1/r)}{\partial n} - \frac{1}{r}\frac{\partial \varphi}{\partial n} \right) \mathrm{d}f_{P'} = 4\pi \int \frac{\varrho_{P'}}{r_{PP'}} \mathrm{d}V_P \,.$$

Now, the integral over the sphere K vanishes because of our assumption about the behavior of φ at infinity. In addition, the second term in the K_P integral does not contribute. Since $\dfrac{\partial(1/r)}{\partial n} = \dfrac{1}{r^2}$ on the surface of the small sphere, the K_P integral yields $4\pi\varphi_P$. Hence,

$$\varphi_P = \int \frac{\varrho_{P'}}{r_{PP'}} \mathrm{d}V_{P'} \,,$$

which is Coulomb's law.

If surface charges are present, the proof proceeds in a similar manner. However, the proof is not so simple for the case of point charges.

7. DIELECTRICS, PHENOMENOLOGICAL TREATMENT

We consider a charged condenser in vacuum formed by two parallel metal plates. We will assume that the areas of the plates are large compared with their separation so that boundary effects need not be considered. Let the surface charge density be $+\omega$ on one plate and $-\omega$ on

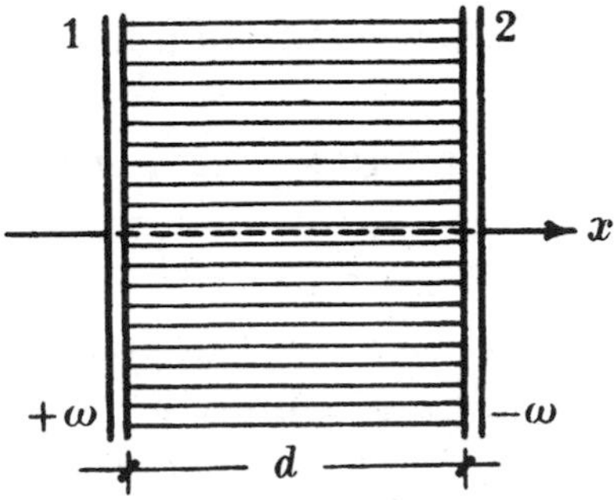

Figure 7.1

the other (see Fig. 7.1). Under these circumstances a uniform electric field exists between the plates,

$$E_x = +4\pi\omega\,,$$

while the field outside is zero. The potential difference between the two plates is thus

$$\varphi_1 - \varphi_2 = 4\pi\omega d\,,$$

where d is the plate separation.

If one brings an *insulator* between the plates, then the following takes place:

1. If the plates are connected to a battery so that *their potential remains constant*, a current flows. *Their charge increases.*
2. If, on the other hand, the plates are insulated, then *their charge remains constant* while *their potential difference decreases* by a factor ε:

$$\varphi_1 - \varphi_2 = \frac{4\pi\omega}{\varepsilon}\,d\,.$$

The factor ε is called the *dielectric constant* of the insulator and is, in general, greater than unity [A-2].

The above phenomenon is interpreted as follows: A layer of charge of magnitude $-(1-1/\varepsilon)$ on the side near plate 1 and $+(1-1/\varepsilon)$ on the side near plate 2 is formed on the outer surfaces of the dielectric. In conjunction with the charge on the plates, this results in a net charge of ω/ε at 1 and $-\omega/\varepsilon$ at 2.

The charge formed on the surfaces of the dielectric cannot be directly observed. Its existence can be demonstrated only through the potential difference. It cannot be altered by means of electrical conduction. Instead, it adheres to the dielectric substance and simply adjusts itself according to the external field. This type of charge is called *polarization charge* to differentiate it from the *true* (*conduction*) *charge* which exists on the metal plates. The latter charge can be altered at will.

In what follows, we will use ϱ_p and ω_p to denote polarization charge densities, ϱ_t and ω_t for true charge densities. Thus $\varrho=\varrho_p+\varrho_t$ and $\omega=\omega_p+\omega_t$ represent total charge densities.

If one imagines a thin region of vacuum to exist between

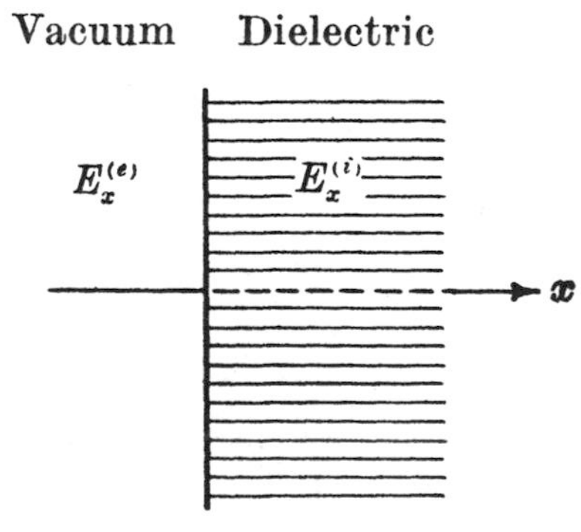

Figure 7.2

the condenser plates and the dielectric, then the relationships between the charge densities and the fields within the condenser are (see Fig. 7.2):

$$E_x^{(e)} = 4\pi\omega\,,$$
$$E_x^{(i)} - E_x^{(e)} = 4\pi\omega_p\,, \quad \text{where} \quad \omega_p = -\left(1 - \frac{1}{\varepsilon}\right)\omega\,.$$

Hence

$$E_x^{(i)} = 4\pi\left[\omega - \left(1 - \frac{1}{\varepsilon}\right)\omega\right] = \frac{4\pi}{\varepsilon}\omega\,,$$

or

$$E_x^{(i)} = \frac{1}{\varepsilon}E_x^{(e)}\,,$$

and

$$4\pi\omega_p = -(\varepsilon - 1)E_x^{(i)}\,.$$

Thus, at the bounding surfaces between the dielectric and

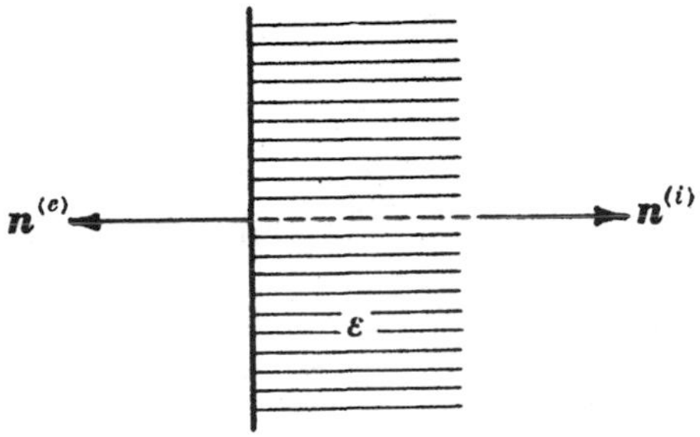

Figure 7.3

the vacuum, in the absence of a true charge density,

$$E_n^{(e)} + \varepsilon E_n^{(i)} = 0\,,$$
$$E_n^{(e)} + E_n^{(i)} = 4\pi\omega_p = -(\varepsilon - 1)E_n^{(i)}\,.$$

This relation can easily be generalized to apply to the

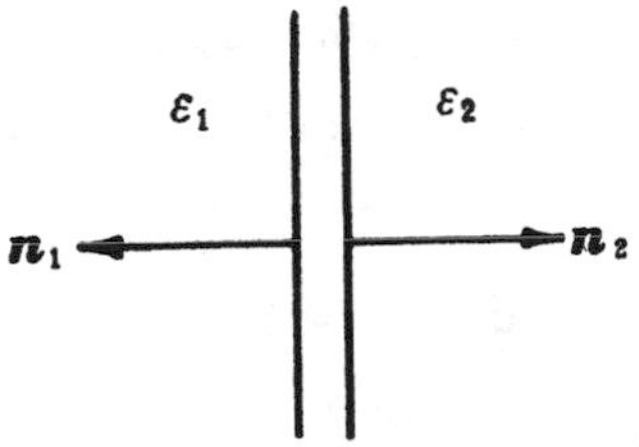

Figure 7.4

bounding surface between two arbitrary dielectrics by imagining a thin vacuum region to exist between them (Fig. 7.4). In this case, in the absence of true charge,

$$\varepsilon_1 E_{n_1} + \varepsilon_2 E_{n_2} = 0 .$$

If there is true charge present,

$$\varepsilon_1 E_{n_1} + \varepsilon_2 E_{n_2} = 4\pi\omega_t .$$

For vacuum, ε is equal to unity.

If we introduce a new vector, the ***displacement D***, defined by

$$\boldsymbol{D} = \varepsilon \boldsymbol{E} , \tag{7.1}$$

then the most general relation for the bounding surface between two insulators becomes

$$D_{n_1} + D_{n_2} = 4\pi\omega_t . \tag{7.2}$$

Since

$$E_{n_1} + E_{n_2} = 4\pi\omega \tag{7.3}$$

(where $\omega = \omega_p + \omega_t$) is still valid, it follows that

$$(\varepsilon_1 - 1)E_{n_1} + (\varepsilon_2 - 1)E_{n_2} = -4\pi\omega_p . \tag{7.4}$$

This is an important property of matter.

In the uniform field of the parallel plate condenser *polarization charges are produced only at the surfaces of the dielectric.* However, in the most general case, a *volume density of polarization charge* ϱ_p is produced within an insulator.

Since the ω_t's are the surface sources of the field $\boldsymbol{D}$, then, by means of a limiting process (a series of stacked sheets, each with true surface charge density ω_t), it can be seen that the ϱ_t's, the volume densities of true charge, are the volume sources of $\boldsymbol{D}$:

$$\operatorname{div} \boldsymbol{D} = 4\pi\varrho_t . \tag{7.5}$$

Because

$$\operatorname{div} \boldsymbol{E} = 4\pi\varrho \tag{7.6}$$

(where $\varrho = \varrho_p + \varrho_t$), it also follows that

$$\operatorname{div}\{(\varepsilon - 1)\boldsymbol{E}\} = -4\pi\varrho_p . \qquad [7.7]$$

If the medium is homogeneous, that is, if ε is independent of position, then

$$(\varepsilon - 1)\operatorname{div}\boldsymbol{E} = -4\pi\varrho_p . \qquad [7.8]$$

If we apply Gauss's theorem to the vector $(\varepsilon - 1)\boldsymbol{E}$ in the usual manner (excluding, however, the surfaces with $\omega_p \neq 0$), then we see that

$$\int \varrho_p \,\mathrm{d}V + \int \omega_p \,\mathrm{d}f = 0. \qquad [7.9]$$

Thus, the sum of all polarization charges is zero.

If $\omega_t = 0$ on a boundary layer between two dielectric media, then D_n, the normal component of $\boldsymbol{D}$, is continuous, while $D_\parallel$, the parallel component, is, in general, not. This results from the fact that $E_\parallel$ is continuous, while $\boldsymbol{E} = -\operatorname{grad}\varphi$, with φ continuous, is still valid just as before.

8. ELECTRON THEORY INTERPRETATION OF DIELECTRIC PHENOMENA

In electron theory the polarization charges, introduced so far on purely phenomenological grounds, are traced back to *electric dipoles.*

The electric dipole

We first consider two point charges $+e$ and $-e$ separated by a distance $\boldsymbol{d}$. The *electric dipole moment* of this arrange-

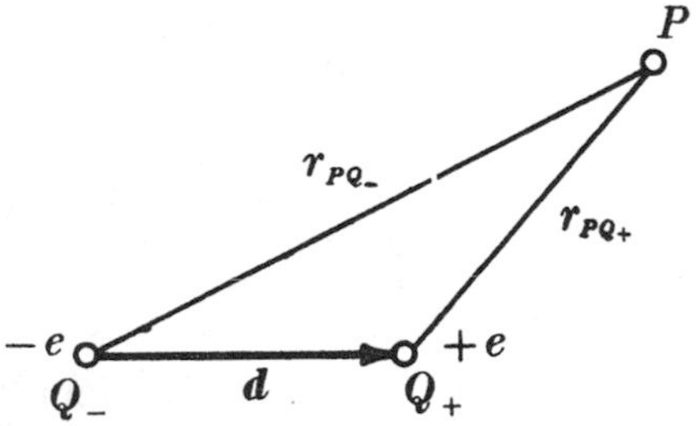

Figure 8.1

ment is defined by the vector

$$\boldsymbol{p} = e\boldsymbol{d}\,.$$

(In general, for an arbitrary number of charges whose sum is zero, one defines

$$\boldsymbol{p} = \sum_i e_i \boldsymbol{x}_i\,,$$

where the $\boldsymbol{x}_i$ are the position vectors of the charges. Since $\sum_i e_i = 0$, this vector is independent of the position of the origin of the coordinate system.)

The potential due to the two charges is, according to Fig. 8.1,

$$\varphi_P = \frac{e}{r_{PQ+}} - \frac{e}{r_{PQ-}}\,.$$

Now, if $r \gg d$ (i.e., either $r \to \infty$, or else $e \to \infty$ as $\boldsymbol{d} \to 0$ so that ed remains finite), this arrangement of two charges is called a dipole. A Taylor expansion of the potential due to this dipole yields

$$\varphi_P = e(\boldsymbol{d}\cdot \operatorname{grad}_Q)\frac{1}{r_{PQ}} = (\boldsymbol{p}\cdot \operatorname{grad}_Q)\frac{1}{r_{PQ}}\,. \qquad [8.1]$$

(Here, grad_Q denotes the gradient taken with respect to the coordinates of Q.) The field produced by the dipole is thus

$$\boldsymbol{E}_P = (\boldsymbol{p}\cdot \operatorname{grad}_Q)\frac{\boldsymbol{x}_P - \boldsymbol{x}_Q}{r_{PQ}^3}\,. \qquad [8.2]$$

Written in component notation,

$$E_k = \left(\sum_i p_i \frac{\partial}{\partial x_{Qi}}\right)\frac{x_{Pk} - x_{Qk}}{r_{PQ}^3} \qquad (i, k = 1, 2, 3)\,.$$

We now consider a finite piece of an insulator. Let there be a dipole present at each atom or molecule. (The question of how these dipoles originate will, for the present, be avoided.) The field produced by these dipoles is then

$$\boldsymbol{E}_P = \sum_Q (\boldsymbol{p} \cdot \mathrm{grad}_Q) \frac{\boldsymbol{x}_P - \boldsymbol{x}_Q}{r_{PQ}^3} . \qquad [8.3]$$

We wish to transform this sum into an integral by making the approximation of replacing the discontinuous distribution of dipoles by a continuous one. Let $\boldsymbol{p}$ be the electric dipole moment per molecule and N the number of molecules per cubic centimeter. Then, the dipole moment per cubic centimeter, $\boldsymbol{P} = N\boldsymbol{p}$, is called the *polarization*. The field then becomes

$$\boldsymbol{E}_P = \int (\boldsymbol{P} \cdot \mathrm{grad}_Q) \frac{\boldsymbol{x}_P - \boldsymbol{x}_Q}{r_{PQ}^3} \, \mathrm{d}V_Q . \qquad [8.4]$$

We wish to transform this integral. For our purposes we will need a generalization of Gauss's theorem: For arbitrary vectors

$$(\boldsymbol{A} \cdot \mathrm{grad})\boldsymbol{B} + (\mathrm{div}\, \boldsymbol{A})\boldsymbol{B} = \mathrm{div}\, (\boldsymbol{B}, \boldsymbol{A}) , \qquad [8.5]$$

where the divergence on the right-hand side is contracted with $\boldsymbol{A}$. In component notation

$$\sum_i \left(A_i \frac{\partial}{\partial x_i} \right) B_k + \left(\sum_i \frac{\partial A_i}{\partial x_i} \right) B_k = \sum_i \frac{\partial}{\partial x_i} (B_k A_i) . \qquad [8.6]$$

For regular vector functions then, according to Gauss's theorem,

$$\int (\boldsymbol{A} \cdot \mathrm{grad})\boldsymbol{B} \, \mathrm{d}V = -\int (\mathrm{div}\, \boldsymbol{A})\boldsymbol{B} \, \mathrm{d}V + \oint A_n \boldsymbol{B} \, \mathrm{d}f . \qquad [8.7]$$

In accordance with this relation, we obtain

$$\boldsymbol{E}_P = -\int (\mathrm{div}\, \boldsymbol{P}) \frac{\boldsymbol{x}_P - \boldsymbol{x}_Q}{r_{PQ}^3} \, \mathrm{d}V - \oint P_n^{(i)} \frac{\boldsymbol{x}_P - \boldsymbol{x}_Q}{r_{PQ}^3} \, \mathrm{d}f .$$

Here, $P_n^{(i)}$ represents the inwardly-directed normal component of $\boldsymbol{P}$. This form for $\boldsymbol{E}_P$ shows that *our continuous distribution of polarization is completely equivalent to a volume and surface distribution of polarization charge.* Indeed, from the first integral

$$\varrho_p = -\operatorname{div} \boldsymbol{P}, \tag{8.8}$$

while the second integral yields

$$\omega_p = -P_n^{(i)}$$

for the polarization surface charge density, provided that the region of integration lies only to one side of the surface so that the surface bounds the piece of insulator under consideration. On the other hand, the surface may be discontinuous and located within the insulator in such a manner that it is enclosed on both sides by the region of integration. The total polarization charge density on such a surface is given by

$$\omega_p = -(P_{n_1} + P_{n_2}). \tag{8.9}$$

This second case is the more general one and includes the first as a special case ($P_{n_2} = 0$).

Gauss's theorem applied to the vector $\boldsymbol{P}$ shows once again that

$$\int \varrho_p \, \mathrm{d}V + \int \omega_p \, \mathrm{d}f = 0. \tag{8.10}$$

Furthermore, it is seen that in order to preserve the phenomenological laws of the previous section, $\boldsymbol{P}$ must be identified with $(\varepsilon - 1)\boldsymbol{E}/4\pi$. Therefore,

$$\boldsymbol{D} = \boldsymbol{E} + 4\pi \boldsymbol{P}. \tag{8.11}$$

The dielectric constant ε can also be expressed in terms of atomic quantities by making the obvious assumption

$$\boldsymbol{p} = \alpha \boldsymbol{E}', \tag{8.12}$$

in which α is the molecular polarizability and $\boldsymbol{E}'$ is the effective field; i.e., the field prevailing at the position of the dipole $\boldsymbol{p}$ but not including the field produced by the dipole itself. If we make the rough approximation that

$\boldsymbol{E}'=\boldsymbol{E}$, then

$$\boldsymbol{P}=N\alpha\boldsymbol{E}$$

and

$$\varepsilon=1+4\pi\alpha N\,. \qquad [8.13]$$

Empirically, it is found that $\alpha>0$ for electrostatic fields. Consequently, ε is greater than one [A-2].

We now wish to refine this theory by considering the difference between $\boldsymbol{E}'$ and $\boldsymbol{E}$. The macroscopic field is just the average of the microscopic field:

$$\boldsymbol{E}=\frac{1}{V}\int\boldsymbol{E}_{\text{mol}}\,\mathrm{d}V\,.$$

If we denote the field of a single dipole by $\boldsymbol{e}$, then

$$\boldsymbol{E}=\boldsymbol{E}'+N\int_K\boldsymbol{e}\cdot\mathrm{d}V\,.$$

Here, we have assumed that the medium is isotropic. The integral extends over a sphere that contains the dipole. We will see that it is independent of the size of the sphere.

Calculation of $\int_K \boldsymbol{e}\,\mathrm{d}V$.[2] The geometry is shown in Fig. 8.2. Cylindrical coordinates ϱ, ζ will be used. We represent the

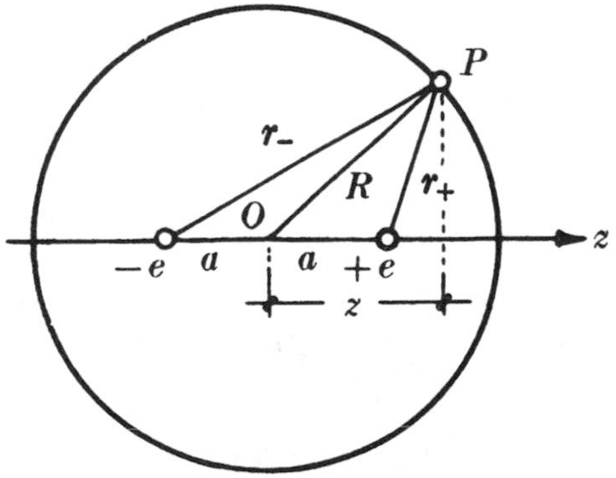

Figure 8.2

[2] Lundblad, *Ann. Physik* **57**, 183 (1918).

radius of the sphere by R and $P(\varrho, z)$ is a point on its surface. Clearly,

$$R^2 = \varrho^2 + z^2 .$$

From symmetry considerations, only the z component of

$$\boldsymbol{e} = -e \cdot \operatorname{grad}\left(\frac{1}{r_+} - \frac{1}{r_-}\right)$$

can contribute. Since

$$e_z = -e \frac{\partial}{\partial \zeta}\left\{\frac{1}{\sqrt{\varrho^2 + (\zeta - a)^2}} - \frac{1}{\sqrt{\varrho^2 + (\zeta + a)^2}}\right\},$$

the integral becomes

$$\int_K e_z \,\mathrm{d}V = -e \int_0^R 2\pi\varrho \,\mathrm{d}\varrho \int_{-z=-\sqrt{R^2-\varrho^2}}^{+z=+\sqrt{R^2-\varrho^2}} \mathrm{d}\zeta \frac{\partial}{\partial \zeta}\left\{\frac{1}{\sqrt{\varrho^2 + (\zeta - a)^2}} - \frac{1}{\sqrt{\varrho^2 + (\zeta + a)^2}}\right\},$$

$$= -e \int_0^R 4\pi\varrho \,\mathrm{d}\varrho \left[\frac{1}{\sqrt{\varrho^2 + (z - a)^2}} - \frac{1}{\sqrt{\varrho^2 + (z + a)^2}}\right],$$

$$= -e \int_0^R 4\pi z \,\mathrm{d}z \left[\frac{1}{\sqrt{R^2 + a^2 - 2az}} - \frac{1}{\sqrt{R^2 + a^2 + 2az}}\right],$$

$$= -e \int_{-R}^{+R} \frac{4\pi z \,\mathrm{d}z}{\sqrt{R^2 + a^2 - 2az}} .$$

With the substitution $u^2 = R^2 + a^2 - 2az$ one obtains

$$\int_K e_z \mathrm{d}V = -e \cdot 4\pi \int_{|R-a|}^{R+a} \frac{R^2 + a^2 - u^2}{2a^2} \,\mathrm{d}u$$

$$= -4\pi e \cdot \frac{(R^2 + a^2)u - \frac{1}{3}u^3}{2a^2}\Bigg|_{|R-a|}^{R+a} .$$

If we now assume that $R > a$, then this expression becomes, in fact, independent of R:

$$\int_K e_z \,\mathrm{d}V = -\frac{4\pi}{3} e\, 2a = -\frac{4\pi}{3} p\,.$$

Here, $p = 2ea$ is the moment of the dipole. Thus, as the contribution of the field of a dipole to the macroscopic field, we obtain

$$\int_K \boldsymbol{e}\,\mathrm{d}V = -\frac{4\pi}{3}\boldsymbol{p}\,.$$

It is essential that the above calculation be performed with a finite dipole ($a \not\to 0$), since otherwise the integrand would diverge when expanded in powers of a.

For *isotropic media* there results, then,

$$\boldsymbol{E} = \boldsymbol{E}' - \frac{4\pi}{3} N\boldsymbol{p} = \boldsymbol{E}' - \frac{4\pi}{3}\boldsymbol{P}\,. \qquad [8.14]$$

In general, this formula is not valid for anisotropic media.

As a consequence of

$$\boldsymbol{P} = \frac{1}{4\pi}(\varepsilon - 1)\boldsymbol{E}$$

and

$$\boldsymbol{P} = N\alpha\boldsymbol{E}' = N\alpha\left(\boldsymbol{E} + \frac{4\pi}{3}\boldsymbol{P}\right)$$

$$= N\alpha\left[1 + \frac{1}{3}(\varepsilon - 1)\right]\boldsymbol{E}\,,$$

the relation between the molecular polarizability and the dielectric constant is given by

$$\frac{\varepsilon - 1}{\varepsilon + 2} = \frac{4\pi}{3} N\alpha\,. \qquad [8.15]$$

This is the *Clausius-Mossotti* relation for isotropic media.

If $\varepsilon - 1 \ll 1$, then $\varepsilon + 2 \approx 3$ and

$$\varepsilon - 1 \approx 4\pi N\alpha ,$$

and we thus obtain the approximate formula of Eq. [8.13].

9. THE POTENTIAL PROBLEM

On the basis of Coulomb's law, we are in a position to calculate the potential due to a given charge distribution. However, in practice, other quantities are usually given and the charge distribution is ordinarily only implicitly determined.

The conditions which φ must satisfy are as follows:

1. At the *boundary surfaces of conductors*, either

(*a*) $-\oint \varepsilon(\partial\varphi/\partial n)\mathrm{d}f = 4\pi e$ for an isolated conductor with the total charge specified. In addition, it is known that $\varphi = \text{const.}$ (The value of the constant is, however, not given in this case); or

(*b*) $\varphi = \varphi_0$ if the conductor is connected to a source of constant potential. In this case the potential is given but not the charge.

2. At the *boundary surfaces of insulators*, if $\omega_t = 0$, φ and $\varepsilon\partial\varphi/\partial n$ are continuous.

These are the *boundary conditions* for φ. In addition, there is also the differential equation

$$\operatorname{div}(\varepsilon \operatorname{grad}\varphi) = 0, \qquad [9.1]$$

which is valid for *vacuum* and for *insulators* where ε is continuous. The region inside conductors is of no interest since we know that φ is constant there. In vacuum where $\varepsilon = 1$ or, in general, where $\varepsilon = \text{constant}$, this differential equation becomes

$$\nabla^2\varphi = 0 . \qquad [9.2]$$

It can be shown that under these conditions a solution

exists. The proof will not be given here—physically, the existence of a solution is obvious. In addition, it can easily be shown that the solution is unique provided that $\varepsilon > 0$ (which is physically obvious [A-2]) and that φ at infinity vanishes at least as fast as $1/r$.

Proof of the uniqueness of the solution. Assumption: Two solutions φ_1 and φ_2 exist, each satisfying all the required conditions. We then consider the difference

$$\psi = \varphi_1 - \varphi_2 .$$

At conducting surfaces, either

(*a*) $\oint \varepsilon(\partial\psi/\partial n)\,\mathrm{d}f = 0, \qquad \psi = \text{constant}$

or

(*b*) $\psi = 0,$

that is, $\oint \psi\varepsilon(\partial\psi/\partial n)\,\mathrm{d}f = 0$ in both cases.

At the boundary surfaces of insulators, $\varepsilon\partial\psi/\partial n$ and ψ are continuous, and $\operatorname{div}(\varepsilon \operatorname{grad}\psi) = 0$ over the whole region where ε is continuous. Because of

$$\operatorname{div}(\psi\varepsilon \operatorname{grad}\psi) \equiv \psi \operatorname{div}(\varepsilon \operatorname{grad}\psi) + \varepsilon(\operatorname{grad}\psi)^2 ,$$

Gauss's theorem, applied to the inside of a very large sphere from which all metallic bodies and discontinuous surfaces are excluded, yields

$$\underset{\substack{K\\\text{large}\\\text{sphere}}}{\oint} \psi\varepsilon \frac{\partial\psi}{\partial n}\,\mathrm{d}f + \underset{\substack{\text{conductor}\\\text{surfaces}}}{\oint} \psi\varepsilon \frac{\partial\psi}{\partial n}\,\mathrm{d}f - \underset{\substack{\text{insulator}\\\text{surfaces}}}{\oint} \psi\left(\varepsilon_1 \frac{\partial\psi}{\partial n_1} + \varepsilon_2 \frac{\partial\psi}{\partial n_2}\right)\mathrm{d}f$$

$$= \int \varepsilon(\operatorname{grad}\psi)^2\,\mathrm{d}V = 0 ,$$

since all three integrals on the left vanish (the first because ψ vanishes as $1/r$ as $r \to \infty$). Because $\varepsilon > 0$, the integrand is positive definite. It can vanish only if $\operatorname{grad}\psi = 0$ and $\psi = \text{const}$. However, since $\psi = 0$ at infinity then $\psi = 0$ everywhere, thereby proving the uniqueness of the solution.

Mathematical remark. The differential equation

$$\operatorname{div}(\varepsilon \operatorname{grad}\varphi) = 0$$

is equivalent to the variational problem

$$J = \frac{1}{8\pi}\int \varepsilon(\operatorname{grad}\varphi)^2 \,\mathrm{d}V - \sum_i e_i\varphi_i = \begin{matrix}\text{minimum for given}\\ \text{boundary conditions.}\end{matrix}$$

The sum over i is to extend over all conductors on which the charge is specified (Case *a*). Thus,

$$\delta J = \frac{1}{4\pi}\int \varepsilon \operatorname{grad}\varphi\cdot\operatorname{grad}(\delta\varphi)\,\mathrm{d}V - \sum_i e_i\,\delta\varphi_i\,,$$

$$= -\frac{1}{4\pi}\int \operatorname{div}(\varepsilon\operatorname{grad}\varphi)\,\delta\varphi\,\mathrm{d}V + \frac{1}{4\pi}\int \varepsilon\frac{\partial\varphi}{\partial n}\,\delta\varphi\,\mathrm{d}f - \sum_i e_i\,\delta\varphi_i = 0\,.$$

The surface integral vanishes on all surfaces except those of conductors of Case *a*. For such surfaces it yields $\sum_i e_i\delta\varphi_i$ since n is the inward normal, and there remains

$$\delta J = -\frac{1}{4\pi}\int \operatorname{div}(\varepsilon\operatorname{grad}\varphi)\,\delta\varphi\,\mathrm{d}V = 0\,.$$

The differential equation [9.1] is, therefore, a necessary condition.

On the other hand, it can be shown that because $\varepsilon > 0$, then, together with the boundary conditions, it is also sufficient for a minimum. Indeed, let φ satisfy the differential equation and the boundary conditions. As the general solution of the minimal problem we take

$$\bar{\varphi} = \varphi + \varphi'\,.$$

We then obtain

$$J(\bar{\varphi}) = J(\varphi) + \frac{1}{8\pi}\int \varepsilon(\operatorname{grad}\varphi')^2\,\mathrm{d}V\,,$$

$$J(\bar{\varphi}) \geqslant J(\varphi)\,,$$

so that φ is indeed a solution of the minimal problem.

Physical interpretation. The significance of the integrand which occurs in the variational problem is that of an *energy density*. Indeed,

$$\frac{1}{8\pi}\,\varepsilon(\operatorname{grad}\varphi)^2 = \frac{1}{8\pi}\,\varepsilon \boldsymbol{E}^2 = \frac{1}{8\pi}\,(\boldsymbol{E}\cdot\boldsymbol{D}) = \frac{1}{8\pi}\,\boldsymbol{E}^2 + \frac{1}{2}\,(\boldsymbol{E}\cdot\boldsymbol{P})\,.$$

Here $\boldsymbol{E}^2/8\pi$ is the energy density of the field in vacuum and $(\boldsymbol{E}\cdot\boldsymbol{P})/2$ is the potential energy per unit volume of the polarization. (The factor of $\frac{1}{2}$ arises from the fact that the polarization is proportional to $\boldsymbol{E}$.) The sum of these two thus gives the energy density in a dielectric.

10. CURVILINEAR COORDINATES

Expressed in *orthogonal curvilinear coordinates* u, v, and w, the square of the line element has the form

$$\mathrm{d}s^2 = \sum_{i=1}^{3} (\mathrm{d}x_i)^2 = e_1^2\,\mathrm{d}u^2 + e_2^2\,\mathrm{d}v^2 + e_3^2\,\mathrm{d}w^2\,. \qquad [10.1]$$

(There are no cross terms in the case of orthogonal coordinates.) Here, e_1, e_2, and e_3 are functions of u, v, and w.

The scalar product of a vector $\boldsymbol{A}$ with the line element is

$$A_s\,\mathrm{d}s = \boldsymbol{A}\cdot\mathrm{d}\boldsymbol{x} \equiv A_u e_1\,\mathrm{d}u + A_v e_2\,\mathrm{d}v + A_w e_3\,\mathrm{d}w\,.$$

If one sets $\boldsymbol{A} = \operatorname{grad}\varphi$, then the expressions

$$\frac{\partial\varphi}{\partial u}\,\mathrm{d}u + \frac{\partial\varphi}{\partial v}\,\mathrm{d}v + \frac{\partial\varphi}{\partial w}\,\mathrm{d}w$$

and

$$(\operatorname{grad}\varphi)_u e_1\,\mathrm{d}u + (\operatorname{grad}\varphi)_v e_2\,\mathrm{d}v + (\operatorname{grad}\varphi)_w e_3\,\mathrm{d}w$$

must be identical; i.e., the u, v, and w components of $\operatorname{grad}\varphi$ are

$$(\operatorname{grad}\varphi)_u = \frac{1}{e_1}\frac{\partial\varphi}{\partial u}\,,\ (\operatorname{grad}\varphi)_v = \frac{1}{e_2}\frac{\partial\varphi}{\partial v}\,,\ (\operatorname{grad}\varphi)_w = \frac{1}{e_3}\frac{\partial\varphi}{\partial w}\,. \qquad [10.2]$$

The surface element tensor is

$$\mathrm{d}f_{uv} = e_1 e_2 \,\mathrm{d}u\,\mathrm{d}v\ ,\quad \mathrm{d}f_{vw} = e_2 e_3\,\mathrm{d}v\,\mathrm{d}w\ ,\quad \mathrm{d}f_{uw} e_1 e_3\,\mathrm{d}u\,\mathrm{d}w\ ,$$

and the product of the surface element with the component of curl $\boldsymbol{A}$ perpendicular to it is

$$\mathrm{curl}_n \boldsymbol{A}\,\mathrm{d}f = (\mathrm{curl}\,\boldsymbol{A})_u e_2 e_3\,\mathrm{d}v\,\mathrm{d}w + (\mathrm{curl}\,\boldsymbol{A})_v e_1 e_3\,\mathrm{d}u\,\mathrm{d}w + \dots .$$

Employing Stokes's theorem,

$$\oint A_s\,\mathrm{d}s = \int \left[\frac{\partial(e_3 A_w)}{\partial v} - \frac{\partial(e_2 A_v)}{\partial w}\right] \mathrm{d}v\,\mathrm{d}w + \dots ,$$

one obtains, upon comparison,

$$(\mathrm{curl}\,\boldsymbol{A})_u = \frac{1}{e_2 e_3}\left[\frac{\partial(e_3 A_w)}{\partial v} - \frac{\partial(e_2 A_v)}{\partial w}\right],\ \text{etc}\,. \qquad [10.3]$$

It can easily be verified that $\mathrm{curl}\,(\mathrm{grad}\,\varphi) = 0$.

Because

$$A_n\,\mathrm{d}f = A_u e_2 e_3\,\mathrm{d}v\,\mathrm{d}w + A_v e_1 e_3\,\mathrm{d}u\,\mathrm{d}w + A_w e_1 e_2\,\mathrm{d}u\,\mathrm{d}v\ ,$$

Gauss's theorem can be written as

$$\oint A_n\,\mathrm{d}f = \int \left[\frac{\partial}{\partial u}(e_2 e_3 A_u) + \frac{\partial}{\partial v}(e_1 e_3 A_v) + \dots\right] \mathrm{d}u\,\mathrm{d}v\,\mathrm{d}w\,.$$

Since the integrand in the second term is $\mathrm{div}\,\boldsymbol{A}\,\mathrm{d}V$, and since

$$\mathrm{d}V = e_1 e_2 e_3\,\mathrm{d}u\,\mathrm{d}v\,\mathrm{d}w\ ,$$

then

$$\mathrm{div}\,\boldsymbol{A} = \frac{1}{e_1 e_2 e_3}\left[\frac{\partial}{\partial u}(e_2 e_3 A_u) + \frac{\partial}{\partial v}(e_1 e_3 A_v) + \frac{\partial}{\partial w}(e_1 e_2 A_w)\right]. \qquad [10.4]$$

Upon substituting $A = \mathrm{grad}\,\varphi$ in this last equation, the Laplacian operator $\nabla^2 \equiv \mathrm{div}\,\mathrm{grad}$ is obtained. Thus,

$$\nabla^2\varphi = \frac{1}{e_1 e_2 e_3}\left[\frac{\partial}{\partial u}\left(\frac{e_2 e_3}{e_1}\frac{\partial\varphi}{\partial u}\right) + \frac{\partial}{\partial v}\left(\frac{e_1 e_3}{e_2}\frac{\partial\varphi}{\partial v}\right) + \frac{\partial}{\partial w}\left(\frac{e_1 e_2}{e_3}\frac{\partial\varphi}{\partial w}\right)\right]. \qquad [10.5]$$

This can also be derived from the variational principle:

$$\int (\operatorname{grad} \varphi)^2 \, dV = \min.$$

11. EXAMPLES OF SOLUTIONS OF THE POTENTIAL PROBLEM

By means of examples, we will illustrate two methods for the solution of the potential problem.

a. Method of images

1. *A point charge e opposite an infinite conducting sheet.* We assume that the sheet is grounded so that its potential is $\varphi = 0$. The method of images consists of assuming a virtual charge $-e$ located at the point which is the image of the position of e mirrored at the sheet. This image charge, together with the charge e, is to determine

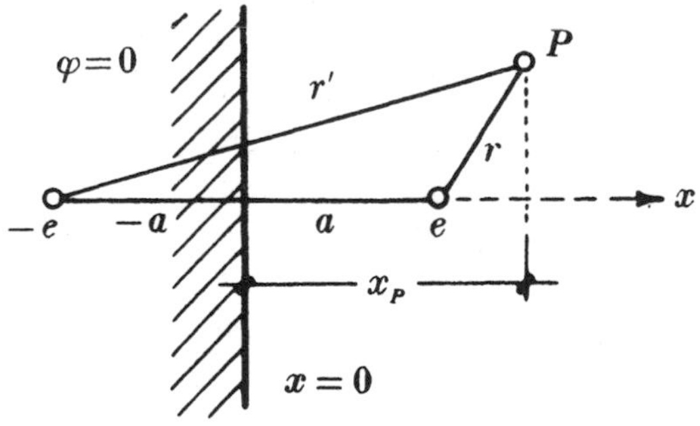

Figure 11.1

the field to the right of the sheet but is to have no effect on the field to the left of the sheet (Fig. 11.1). We thus have

$$x < 0: \quad \varphi = 0\,; \qquad x > 0: \quad \varphi = \frac{e}{r} - \frac{e}{r'}\,.$$

This φ satisfies the equation $\nabla^2 \varphi = 0$ and is continuous for $x = 0$. On the basis of our uniqueness theorem it is, therefore, the solution to the problem.

To find the surface charge induced on the sheet we note that,

$$E_x = \frac{e}{r^3}(x_P - a) - \frac{e}{r'^3}(x_P + a).$$

For $x_P = 0$,

$$r' = r, \qquad E_x = -\frac{2a}{r^3}e = 4\pi\omega.$$

Thus,

$$\omega = -\frac{e}{2\pi}\frac{a}{r^3}.$$

The total charge on the conducting sheet is then

$$\bar{e} = \int_0^\infty \omega 2\pi\varrho \, \mathrm{d}\varrho = \int_a^\infty \omega 2\pi r \, \mathrm{d}r = -e,$$

where $r = \sqrt{\varrho^2 + a^2}$, $r\mathrm{d}r = \varrho \mathrm{d}\varrho$ (ϱ is the distance from the origin to a point on the sheet).

2. *A point charge e opposite a semi-infinite homogeneous dielectric.* In this case the boundary condition is

$$\varepsilon\left(\frac{\partial\varphi}{\partial x}\right)_- = \left(\frac{\partial\varphi}{\partial x}\right)_+,$$

where ε is the dielectric constant of the medium. We again

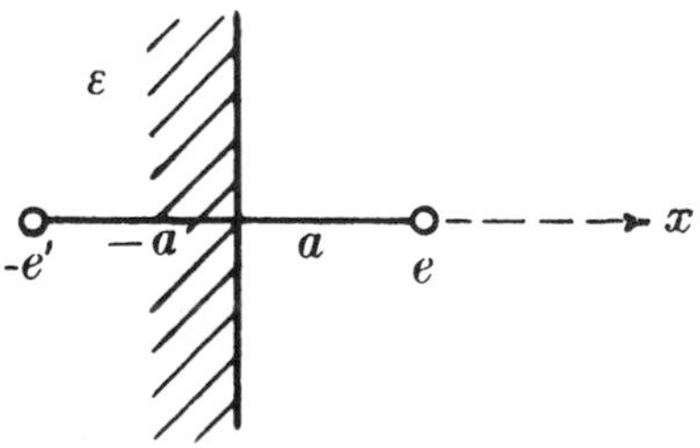

Figure 11.2

assume a mirror charge $-e'$ which contributes to the field to the right of the boundary, but this time the magnitude of e' remains to be determined. Then, for

$$x>0: \quad \varphi=\frac{e}{r}-\frac{e'}{r'},$$

$$x<0: \quad \varphi=\frac{e-e'}{r},$$

so that φ is continuous at $r=r'$. Hence,

$$\left(\frac{\partial\varphi}{\partial x}\right)_+=+\frac{a}{r^3}(e+e'), \qquad \varepsilon\left(\frac{\partial\varphi}{\partial x}\right)_-=+\frac{a}{r^3}\varepsilon(e-e').$$

The boundary condition thus requires that $e+e'=\varepsilon(e-e')$. Thus,

$$e'=\frac{\varepsilon-1}{\varepsilon+1}\cdot e.$$

The solution of our problem is, accordingly,

$$x>0: \quad \varphi=\frac{e}{r}-\frac{\varepsilon-1}{\varepsilon+1}\frac{e}{r'},$$

$$x<0: \quad \varphi=\frac{2}{\varepsilon+1}\frac{e}{r}.$$

3. *A point charge e opposite a conducting sphere.* To every point P can be assigned a "mirror" point P' by means of the relation (Fig. 11.3)

$$\varrho\varrho'=a^2.$$

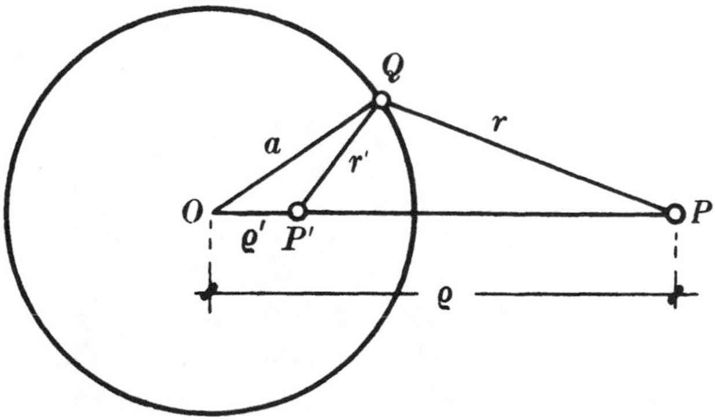

Figure 11.3

Since triangles $OP'Q$ and OQP are similar (they have a common angle at O and $\varrho'/a = a/\varrho$), then

$$r'/r = a/\varrho = \varrho'/a = \sqrt{\varrho'/\varrho} = \text{constant}$$

for all points Q lying on the surface of the sphere. This fact can be used to solve the problem.

Case 1. Grounded sphere. We imagine that a virtual charge $-e'$ at P contributes to the external field. Then,

$$\varphi = \frac{e}{r} - \frac{e'}{r'} \qquad \text{outside the sphere}\,,$$

$$\varphi = 0 \qquad \text{inside the sphere}\,.$$

In order that φ be continuous on the surface of the sphere, the relation

$$e' = e \cdot \frac{r'}{r} = e\sqrt{\frac{\varrho'}{\varrho}}$$

must hold. From this, e' is constant, and the problem is solved. The total charge induced on the sphere is $-e'$ (from Gauss's law).

Case 2. Insulated sphere. We specify that the sphere is uncharged. Thus, the image charge $-e'$ must be balanced, which we do by placing a charge $+e'$ at the center of the sphere. If R represents the distance from the field point to the center of the sphere, then

$$\varphi = \frac{e}{r} - \frac{e'}{r'} + \frac{e'}{R} \qquad \text{for} \quad R > a\,,$$

$$\varphi = \frac{e'}{a} \qquad \text{for} \quad R < a\,,$$

where e' is still given by the formula for Case 1 above.

For the case of the dielectric sphere, the solution cannot be given in closed form; instead, an infinite series is obtained. The same is true for two conducting spheres.

4. *Conducting sphere in a uniform electric field.* This problem can be considered as a limiting condition of Case 2 of 3 above, if $\varrho \to \infty$ and $e \to \infty$ in such a manner that $e/\varrho^2 \to E$.

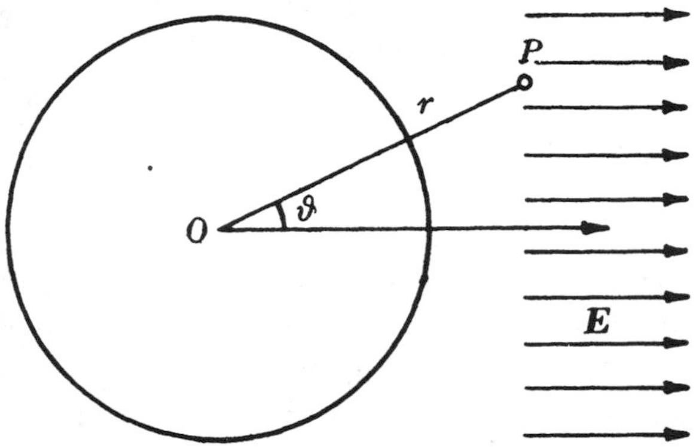

Figure 11.4

It is seen that a dipole of moment p is produced at the center of the sphere:

$$p = e'\varrho' .$$

One can also begin with the following ad hoc hypothesis:

$$r > a: \quad \varphi = -Ex + A + B\frac{x}{r^3},$$

$$r < a: \quad \varphi = A,$$

where Bx/r^3 is the potential of a dipole. Defining $x = r\cos\vartheta$, one can also write

$$r > a: \quad \varphi = \cos\vartheta\left(-Er + \frac{B}{r^2}\right) + A .$$

The continuity of φ can be achieved by setting the quantity in the parentheses equal to zero when $r = a$. Thus,

$$B = Ea^3 .$$

The constant A remains undetermined; it has, however, no physical significance.

The charge distribution induced on the sphere is

$$4\pi\omega = -\left(\frac{\partial\varphi}{\partial r}\right)_{r=a} = \cos\vartheta\left(E + \frac{2B}{a^3}\right) = 3E\cos\vartheta\,.$$

The total charge on the sphere is

$$\oint \omega\, \mathrm{d}f = 0\,, \qquad \text{since} \qquad \oint_{\text{sphere}} \cos\vartheta\, \mathrm{d}f = 0\,.$$

5. *A dielectric sphere embedded in a dielectric medium.* Let the dielectric constant of the sphere be ε_i (=constant) and that of the surrounding medium be ε_a (=constant). At large distances from the sphere let the field be uniform and equal to $\boldsymbol{E}$. We make the hypothesis that for

$$r > a: \qquad \varphi_a = -E\left(x + \varkappa\frac{\partial}{\partial x}\frac{1}{r}\right) = -E\cos\vartheta\left(r - \frac{\varkappa}{r^2}\right),$$

$$r < a: \qquad \varphi_i = -Dx = -Dr\cos\vartheta\,.$$

The boundary conditions at $r=a$ are

$$\varphi_i = \varphi_a: \qquad E\left(1 - \frac{\varkappa}{a^3}\right) = D\,,$$

$$\varepsilon_i\frac{\partial\varphi_i}{\partial r} = \varepsilon_a\frac{\partial\varphi_a}{\partial r}: \qquad E\left(1 + \frac{2\varkappa}{a^3}\right)\varepsilon_a = D\varepsilon_i\,.$$

From this, it follows that the problem is solved if we take

$$\varkappa = \frac{\varepsilon_i - \varepsilon_a}{\varepsilon_i + 2\varepsilon_a}a^3\,, \qquad D = \frac{3\varepsilon_a}{\varepsilon_i + 2\varepsilon_a}E\,.$$

It is interesting to note that for $\varepsilon_i = \varepsilon$ and $\varepsilon_a = 1$, an expression similar to the Clausius-Mossotti relation of Eq. [8.15] is obtained, namely

$$\varkappa = \frac{\varepsilon - 1}{\varepsilon + 2}a^3\,.$$

Because of this, it was possible for Clausius and Mossotti to explain dielectric phenomena by assuming that a dielectric consists of small conducting spheres.[3]

b. Application of function theory to two-dimensional problems

If a spatial charge distribution depends only upon two of the cartesian coordinates, say x and y, so that the charge distribution is the same in all planes perpendicular to the third coordinate axis, then the field in all such planes must be identical and one speaks of a two-dimensional problem. In this case the problem can be solved with the aid of the theory of analytic functions.

Let w be an analytic function of the complex variable $z=x+iy$:

$$w(z) = u + iv \qquad (u,\, v \text{ real})\,.$$

Then, because

$$\frac{dw}{dz} = \frac{\partial}{\partial x}(u+iv) = \frac{1}{i}\frac{\partial}{\partial y}(u+iv)\,,$$

we have the Cauchy-Riemann equations:

$$\frac{\partial u}{\partial x} = \frac{\partial v}{\partial y} \qquad \text{and} \qquad \frac{\partial v}{\partial x} = -\frac{\partial u}{\partial y}\,.$$

From these it follows that

$$\nabla^2 u = 0 \qquad \text{and} \qquad \nabla^2 v = 0\,, \qquad \text{where} \qquad \nabla^2 \equiv \frac{\partial^2}{\partial x^2} + \frac{\partial^2}{\partial y^2}\,.$$

That is, u and v satisfy Laplace's equation. We will, therefore, attempt to determine analytic functions so that their real parts u (or imaginary parts v) represent the desired potential. According to Liouville's theorem, a function that is regular everywhere (including infinity) must be constant.

[3] See *Enzyklopädie der mathematischen Wissenschaften*, vol. V2, p. 329, or the original work of Clausius, *Mechanische Wärmetheorie* (Braunschweig, 1887-91), vol. 2, p. 62.

We will thus have to characterize functions according to their singularities.

Since according to the Cauchy-Riemann equations, $\partial u/\partial n = \partial v/\partial s$, we have (Fig. 11.5)

$$\oint \frac{\partial u}{\partial n}\,ds = \oint \frac{\partial v}{\partial s}\,ds = [v] \qquad \text{(change in } v \text{ after one turn).}$$

If we identify the potential φ with u, then

$$\oint \frac{\partial u}{\partial n}\,ds = [v] = -4\pi\omega\,.$$

Here, ω is the charge in a cylinder of unit height. If we assume that the charge is distributed uniformly along

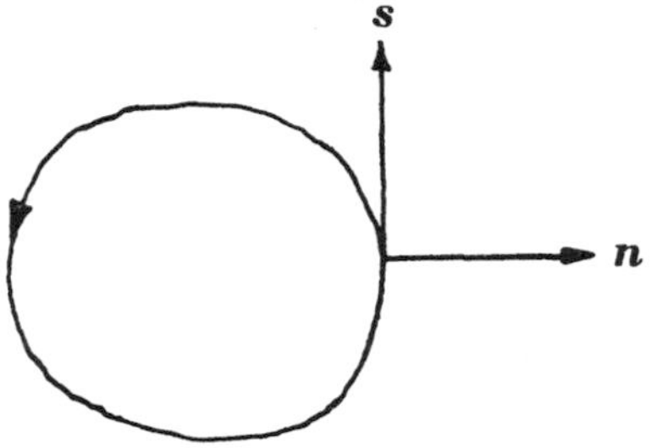

Figure 11.5

a line of infinite extent normal to the x-y plane, then ω represents the charge per unit length.

Now, v is to increase after a complete circuit. This is a property of the logarithm of a complex number; thus $w = \ln(z - z_0)$, v increases by 2π.

We obtain the solution for the case of a *set of line charges perpendicular to the x-y plane* with linear charge densities $\omega_1, \omega_2, \ldots$ and intersecting the x-y plane at the points $z_1, z_2, \ldots$, if we set

$$w = -\sum_Q 2\omega_Q \log(z - z_Q)\,.$$

The potential is the real part of w:

$$\varphi = u_P = -\sum_Q 2\omega_Q \log r_{PQ}\,, \quad \text{where} \quad r_{PQ} = \sqrt{(x_P - x_Q)^2 + (y_P - y_Q)^2}\,.$$

If the projection of the charge distribution on the x-y plane is a set of points, then the potential at these points has a logarithmic singularity.

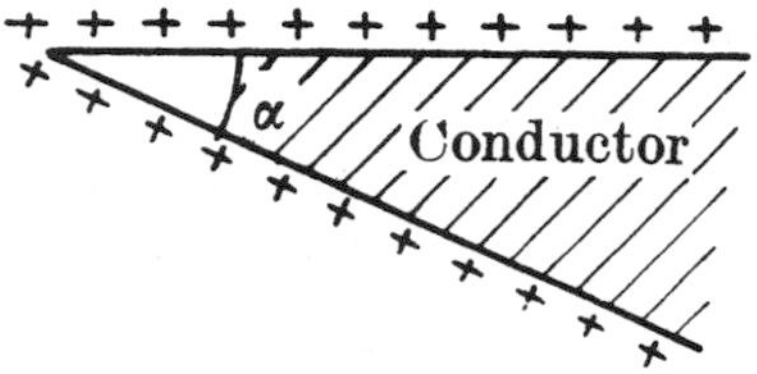

Figure 11.6

As a second example, we will calculate the potential due to a *charged conducting wedge*. Assertion: the potential is given by the imaginary part of the function

$$w = z^\mu,$$

where μ is real and must be suitably determined. For $z = re^{i\vartheta}$ we have

$$\varphi = v = \operatorname{Im} w = r^\mu \sin \mu\vartheta\,.$$

Since the wedge is conducting, we must have

$$v(r, \vartheta = 0) = v(r', \vartheta = 2\pi - \alpha)\,, \qquad \text{all } r,\, r',$$

so that $\mu(2\pi - \alpha) = \pi$. Hence,

$$\mu = \frac{\pi}{2\pi - \alpha}\,.$$

The surface charge density ω on the wedge is given by

$$-4\pi\omega = \frac{1}{r}\frac{\partial v}{\partial \vartheta}\bigg|_{\vartheta=0} = \mu r^{\mu-1} \cos \mu\vartheta \bigg|_{\vartheta=0} = \mu r^{\mu-1}.$$

The total charge $e(r)$ contained within a circle of radius r is

$$-4\pi e(r) = \int_0^{2\pi-\alpha} \frac{\partial v}{\partial r} r \, d\vartheta = -r^\mu \cos \mu\vartheta \Big|_{\vartheta=0}^{\vartheta=2\pi-\alpha} = +2r^\mu$$

12. MAGNETOSTATICS

Electrodynamic phenomena cannot be described by the electric field alone; the magnetic field is also required. Just as electric fields interact with electric charges, magnetic fields interact with so-called magnetic dipoles, with whose aid magnetic fields can then be measured. (Later we will see that magnetic fields can also be measured with moving electric charges.)

A very essential difference as compared with the electric case is that *magnetic monopoles do not exist*; they always occur in pairs so that the total "magnetic charge" is zero. An approximation to a monopole can be produced by removing one member of a pair to a sufficiently large distance (i.e., a very long dipole). It was shown by Gilbert that Coulomb's law is valid for such monopoles.

In analogy to electrostatics we introduce the following vectors:

magnetic field intensity $\boldsymbol{H}$, corresponding to $\boldsymbol{E}$;
magnetization $\boldsymbol{M}$, corresponding to $\boldsymbol{P}$;
magnetic induction $\boldsymbol{B}$, corresponding to $\boldsymbol{D}$.

Aside from the very important difference that in magnetostatics *there is nothing analogous to the true charge densities* ϱ_t, ω_t, the equations for the fields produced by magnetic dipoles correspond to those of electrostatics (Eqs. [2.4], [8.11] and [7.5]):

$$\text{curl } \boldsymbol{H} = 0 \,, \qquad [12.1]$$

$$\boldsymbol{B} = \boldsymbol{H} + 4\pi \boldsymbol{M} \,, \qquad [12.2]$$

$$\text{div } \boldsymbol{B} = 0 \,. \qquad [12.3]$$

At bounding surfaces, the components $H_{\parallel}$ *and* B_n *are continuous.* If we introduce a volume density σ and surface density χ of magnetic "polarization" charge, then, corresponding to Eqs. [8.8] and [8.9],

$$\sigma = -\operatorname{div} \boldsymbol{M} = +\frac{1}{4\pi} \operatorname{div} \boldsymbol{H},$$

$$\chi = -(M_{n_1} + M_{n_2}) = +\frac{1}{4\pi}(H_{n_1} + H_{n_2}).$$

In analogy to Eq. [7.9], it is also true that

$$\int \sigma \, dV + \oint \chi \, df = 0.$$

As in electrostatics (Eq. [7.1]), $\boldsymbol{B}$ and $\boldsymbol{H}$ are related by a constitutive equation,

$$\boldsymbol{B} = \mu \boldsymbol{H}, \qquad [12.4]$$

where μ is the *permeability* and corresponds to ε [A-2]. For paramagnetic substances, μ is greater than unity and is temperature dependent; for diamagnetic substances it is smaller than unity and temperature independent [A-2]. For ferromagnetic substances there is no unique relation between $\boldsymbol{B}$ and $\boldsymbol{H}$. (For example, $\boldsymbol{B}$ and $\boldsymbol{M}$ can be different from zero while $\boldsymbol{H}$ is zero.)

Since $\boldsymbol{H}$ is irrotational, it can be represented by a scalar potential:

$$\boldsymbol{H} = -\operatorname{grad} \psi.$$

Since $\boldsymbol{B}$ is source-free (solenoidal), it can be represented by a vector potential $\boldsymbol{A}$: [4]

$$\boldsymbol{B} = \operatorname{curl} \boldsymbol{A}. \qquad [12.5]$$

However, $\boldsymbol{A}$ is still arbitrary to within the gradient of a

[4] Where there are no magnetizable substances present, the fields $\boldsymbol{H}$ and $\boldsymbol{B}$ are identical and $\boldsymbol{A}$ is then also the vector potential for $\boldsymbol{H}$: $\boldsymbol{H} = \operatorname{curl} \boldsymbol{A}$.

scalar function. Thus, we will even be able to achieve the condition $\operatorname{div} \boldsymbol{A} = 0$.

As an example, we will determine the vector potential of the ***field of a dipole*** of magnetic moment $\boldsymbol{m}$ in the direction of the z axis. As we know from electrostatics, the field produced is

$$\boldsymbol{H} = m \frac{\partial}{\partial z}\left(\operatorname{grad} \frac{1}{r}\right) = m \left(\frac{\partial^2}{\partial x \, \partial z}, \frac{\partial^2}{\partial y \, \partial z}, \frac{\partial^2}{\partial z^2}\right) \frac{1}{r}.$$

Here, differentiation is performed with respect to the coordinates of the field point.

Assertion: the vector potential for $\boldsymbol{H}$ is

$$\boldsymbol{A} = \frac{\boldsymbol{m} \times \boldsymbol{x}}{r^3} = -\boldsymbol{m} \times \operatorname{grad} \frac{1}{r}. \qquad [12.6]$$

Thus,

$$A_x = m \frac{\partial}{\partial y} \frac{1}{r}, \qquad A_y = -m \frac{\partial}{\partial x} \frac{1}{r}, \qquad A_z = 0,$$

so that

$$\operatorname{div} \boldsymbol{A} = 0.$$

For the components of $\boldsymbol{H} = \operatorname{curl} \boldsymbol{A}$ one obtains

$$H_x = m \frac{\partial^2}{\partial x \, \partial z} \frac{1}{r},$$

$$H_y = m \frac{\partial^2}{\partial y \, \partial z} \frac{1}{r},$$

$$H_z = -m \left(\frac{\partial^2}{\partial x^2} + \frac{\partial^2}{\partial y^2}\right) \frac{1}{r} = +m \frac{\partial^2}{\partial z^2} \frac{1}{r},$$

since $\nabla^2(1/r) = 0$. Thus, we have verified that $\boldsymbol{H}$ can be represented by the $\boldsymbol{A}$ given above. Clearly, for a *volume distribution of magnetization*,

$$\boldsymbol{A}_P = \int \mathrm{d}V_Q \frac{\boldsymbol{M} \times \boldsymbol{x}_{QP}}{r_{QP}^3} = -\int \mathrm{d}V_Q \, \boldsymbol{M} \times \operatorname{grad}_P \frac{1}{r_{QP}}, \qquad [12.7]$$

where $\boldsymbol{M}$, representing the magnetic moment per unit volume, is, by definition, the magnetization.

It should now be mentioned that in addition to the previously employed obvious analogy between electric and magnetic quantities,

$\boldsymbol{H}$	$\boldsymbol{E}$
$\boldsymbol{B}$	$\boldsymbol{D}$
$\boldsymbol{M}$	$\boldsymbol{P}$
μ	ε

there exists still another more fundamental one, namely,

$\boldsymbol{H}$	$\boldsymbol{D}$
$\boldsymbol{B}$	$\boldsymbol{E}$
$-\boldsymbol{M}$	$\boldsymbol{P}$
$1/\mu$	ε

as we shall see later.

13. UNITS AND DIMENSIONS

The definition of units and dimensions is, to a certain degree, a matter of convention. The number of fundamental dimensions can be reduced by arranging for certain quantities to be dimensionless. For example, either a new dimension can be chosen for electric charge, or else charge can be referred back to centimeters, grams, and seconds by choosing the constant k in Coulomb's law,

$$K = k\frac{e^2}{r^2},$$

to be dimensionless: $[k]=1$. This definition is arbitrary. (In the analogous case of the gravitational law,

$$F = f\frac{m_1 m_2}{r^2},$$

f is not dimensionless if the dimensions of mass, length, and force are already fixed.) Setting $[k]=1$ corresponds to the cgs system. Thus, since $V = ke^2/r$ represents energy, the dimensions of charge become

$$[\text{charge}] = (\text{erg}\cdot\text{cm})^{\frac{1}{2}} = ([\text{mass}][\text{length}]^3[\text{time}]^{-2})^{\frac{1}{2}} = (ml^3t^{-2})^{\frac{1}{2}}$$

and

$$[\text{field}] = [\text{charge}]\text{cm}^{-2} = (\text{erg}\cdot\text{cm}^{-3})^{\frac{1}{2}} = (ml^{-1}t^{-2})^{\frac{1}{2}} .$$

The *absolute electrostatic charge unit* (*esu*) is defined so that k is dimensionless and equal to one when K and r are measured in cgs units.

On the other hand, if one sets $k = 1/4\pi$, then *Heaviside units* are obtained. In the following, these will be denoted by a subscript H. These units have certain advantages. It follows that

$$K = \frac{e_H^2}{4\pi r^2}, \qquad \text{field intensity} = E_H = \frac{e_H}{4\pi r^2}, \qquad e_H = \sqrt{4\pi} e ,$$

where e is the charge in ordinary esu units,

$$\boldsymbol{E}_H = \frac{1}{\sqrt{4\pi}} \boldsymbol{E}, \quad \boldsymbol{K} = e\boldsymbol{E} = e_H \boldsymbol{E}_H .$$

Then the factor of 4π does not appear in certain other equations:

$$\text{div}\, \boldsymbol{E}_H = \varrho_H , \qquad \text{instead of} \qquad \text{div}\, \boldsymbol{E} = 4\pi\varrho ,$$

while the energy is expressed as

$$W = \frac{1}{8\pi}\int E^2 \,\text{d}V = \frac{1}{2}\int E_H^2 \,\text{d}V .$$

Translator's Note: Another common system is the *rationalized mks* or *Giorgi system*. This system has the advantage of the rationalization (elimination of the factor of 4π in many equations) offered by Heaviside units, and also has the consequence that potential, current, and resistance have the "practical" units of volts, amperes, and ohms.

The fundamental quantities and units in this system are length (meters), mass (kilograms), time (seconds), and charge (coulombs). The *coulomb* is defined as the charge which, on each of two bodies separated by a distance of 1 meter, produces a mutual force of 1 kg·m/sec² ($\equiv$1 newton $=10^5$ dynes).

In the rationalized mks system $k = 1/4\pi\varepsilon_0$, where

$$\varepsilon_0 = 8.454 \times 10^{-12}\ \text{coul}^2\cdot\text{sec}^2\cdot\text{kg}^{-1}\cdot\text{m}^{-3}$$

is called the *permitivity of free space*. Dielectrics are characterized by a permitivity $\varepsilon = \varkappa_e \varepsilon_0$ where $\varkappa_e$ is known as the *relative permitivity* or *dielectric constant* of the material. In magnetostatics the *permeability of free space* is $\mu_0 = 4\pi \times 10^{-7}$ kg·m·coul^{-2}. For magnetic materials, the permeability is written as $\mu = \varkappa_m \mu_0$ in analogy with the electric case.

Chapter 2. Steady-State Currents

14. THEORY OF STEADY-STATE CURRENTS

a. Definitions

The ***current*** J is the electric charge that passes through the total cross-sectional area of a conductor per unit time:

$$J = \frac{\mathrm{d}e}{\mathrm{d}t}. \qquad [14.1]$$

The *current density* $\boldsymbol{i}$ is defined as a vector in the direction of the current flow with magnitude equal to the amount of charge flowing through a unit cross-sectional area perpendicular to $\boldsymbol{i}$ in one second. If the current density in a conductor is independent of position, then

$$|\boldsymbol{i}| = \frac{J}{q},$$

where q is the cross-sectional area of the conductor. Since the current through the surface element $\mathrm{d}f$ in the direction of the normal $\boldsymbol{n}$ is given by $i_n \mathrm{d}f$, then, in the general case,

$$J = \int_q i_n \,\mathrm{d}f. \qquad [14.2]$$

Ordinarily, current is measured in electromagnetic units

(emu). It can, however, also be defined in electrostatic units. In the latter case the dimensions are

$$[J] = [\text{charge}]\text{sec}^{-1} = (\text{erg}\cdot\text{cm}\cdot\text{sec}^{-2})^{\frac{1}{2}} = (ml^3t^{-4})^{\frac{1}{2}},$$

$$[\boldsymbol{i}] = [J]\text{cm}^{-2} = (\text{erg}\cdot\text{cm}^{-3}\cdot\text{sec}^{-2})^{\frac{1}{2}} = (ml^{-1}t^{-4})^{\frac{1}{2}}.$$

b. The continuity equation

Because of the conservation law for electric charge, the amount of charge flowing out of a closed surface per unit time is equal to the decrease per unit time of the charge contained within the surface:

$$\oint_F i_n \, \mathrm{d}f = -\frac{\mathrm{d}e}{\mathrm{d}t} = -\frac{\mathrm{d}}{\mathrm{d}t}\int \varrho \, \mathrm{d}V. \qquad [14.3]$$

If the surface integral is transformed into a volume integral in accordance with Gauss's law, then, since the volume under consideration is arbitrary, it is clear that

$$\frac{\partial \varrho}{\partial t} + \operatorname{div} \boldsymbol{i} = 0. \qquad [14.4]$$

This relation is quite generally valid and is known as the *equation of continuity.*

In the case of *steady-state currents*, all functions are, by definition, time independent: $\partial/\partial t \equiv 0$. Thus, it follows that

$$\operatorname{div} \boldsymbol{i} = 0, \qquad \oint i_n \, \mathrm{d}f = 0. \qquad [14.5]$$

c. Ohm's law

Experimentally, it is found that when a current flows in a material, the following phenomenological relation exists between the vectors $\boldsymbol{i}$ and $\boldsymbol{E}$:

$$\boldsymbol{i} = \sigma \boldsymbol{E}. \qquad [14.6]$$

Here, σ is the *conductivity* of the material in question. The above equation is called *Ohm's law.* It can be explained microscopically if it is assumed that a resistive force proportional to velocity acts on the moving charged particles that constitute the current.

Let N_+ and N_- be the number of particles per unit volume having charges $+e$ and $-e$, respectively. Then,

$$\varrho = (N_+ - N_-)e\,,$$
$$\boldsymbol{i} = N_+ \cdot e \cdot \boldsymbol{v}_+ - N_- \cdot e \cdot \boldsymbol{v}_-,$$

where $\boldsymbol{v}_+$ and $\boldsymbol{v}_-$ are the velocities of the positive and negative particles. Now, Ohm's law implies that each particle is subject to a resistive force

$$\boldsymbol{K}_\pm = -w_\pm \boldsymbol{v}_\pm\,,$$

where w is the specific resistance. The idea is that a stationary state is established where

$$\boldsymbol{K}_\pm = \mp e\boldsymbol{E}\,,$$
$$m\dot{\boldsymbol{v}} = -w_\pm \cdot \boldsymbol{v}_\pm \pm e\boldsymbol{E} = 0\,.$$

In practice, such a stationary state is formed in a very short time. Then,

$$\pm \boldsymbol{v}_\pm = \frac{e}{w_\pm}\boldsymbol{E}\,,$$

from which it follows that

$$\boldsymbol{i} = \sigma \boldsymbol{E}\,, \quad \text{where} \quad \sigma = e^2\left(\frac{N_+}{w_+} + \frac{N_-}{w_-}\right). \qquad [14.7]$$

The first term in σ is practically zero for metals. The frictional force on the electrons in a metal is proportional to their velocity with respect to the positive ion lattice.

The dimensions of conductivity are[1]

$$[\sigma] = \sec^{-1}.$$

Just as before, we have

$$\boldsymbol{E} = -\operatorname{grad}\varphi\,, \qquad \operatorname{curl}\boldsymbol{E} = 0\,,$$

as long as the currents are time independent; that is, it is impossible to gain work from a charged particle which moves around a closed path.

[1] In electrostatic units.

For a *homogeneous conductor* (i.e., a conductor of homogeneous material and of constant cross-sectional area q),

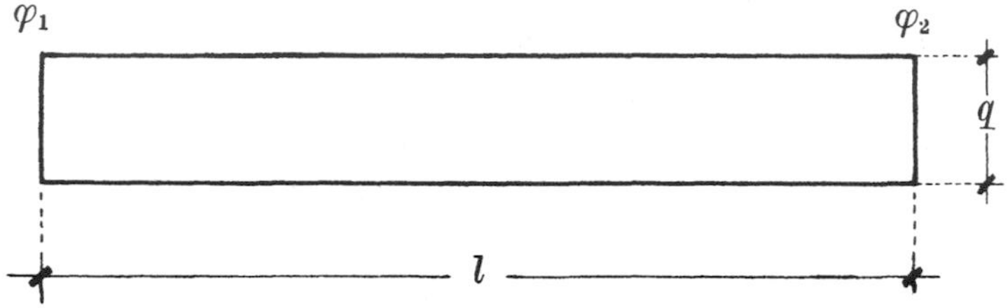

Figure 14.1

Ohm's law takes on the form familiar from experimental physics (see Fig. 14.1):

$$\varphi_1 - \varphi_2 = l\,|\boldsymbol{E}|\,,$$
$$J = q\,|\boldsymbol{i}|,$$

so that

$$\varphi_1 - \varphi_2 = JR\,, \qquad [14.8]$$

where $R = l/(\sigma q)$. The *resistance* R depends upon the physical dimensions of the conductor.

As we shall soon see, Eq. [14.8] is also valid for a conductor of arbitrary shape. In this case, however, R cannot be so simply expressed.

The dimensions of resistance are [2]

$$[R] = [\sigma\,\mathrm{cm}]^{-1} = \mathrm{cm}^{-1} \cdot \mathrm{sec} = 1/\mathrm{velocity}\,.$$

d. *Units*

Current: 1 electromagnetic unit (emu) $= 3 \times 10^{10}$ electrostatic units (esu) $= 10$ amperes;
1 ampere $= 0.1$ emu $= 3 \times 10^{9}$ esu.

Potential: 1 esu $= 3 \times 10^{10}$ emu $= 300$ volts;
1 volt $= 10^{8}$ emu $= 1/300$ esu.

Resistance: 1 ohm $= 1$ volt/1 ampere $= 10^{9}$ emu
$= 1/(9 \times 10^{11})$ esu.

[2] In electrostatic units.

e. Joule's law

Heat is produced wherever electric current flows, and the heat developed per unit volume per unit time is

$$Q = \boldsymbol{i}\cdot\boldsymbol{E} = \sigma E^2 . \qquad [14.9]$$

The dimensions are correct since

$$[\sigma E^2] = \text{erg}\cdot(\text{sec}\cdot\text{cm}^3)^{-1} .$$

This law can also be obtained from microscopic considerations. The work done per unit time by the force $\boldsymbol{K} = e\boldsymbol{E}$ acting on a particle is

$$\boldsymbol{K}\cdot\boldsymbol{v} = e(\boldsymbol{v}\cdot\boldsymbol{E}) .$$

Thus, the work per unit volume and time is

$$e\cdot(N_+\boldsymbol{v}_+ - N_-\boldsymbol{v}_-)\cdot\boldsymbol{E} = \boldsymbol{i}\cdot\boldsymbol{E} .$$

This work expended against resistive forces is completely converted into heat in the conductor. It is called *Joule heat.*

The total heat produced per second in a homogeneous conductor of length l and cross-sectional area q is

$$\bar{Q} = \boldsymbol{i}\cdot\boldsymbol{E}ql = J(\varphi_1 - \varphi_2) = J^2R . \qquad [14.10]$$

We shall see that the expressions $J(\varphi_1 - \varphi_2)$ and J^2R are also correct for arbitrary conductors.

The dimensions of this heat developed per unit time (= power = energy/time) are

$$1 \text{ volt} \times 1 \text{ ampere} = 1 \text{ watt} = 10^7 \text{ erg}\cdot\text{sec}^{-1} = 1 \text{ joule/sec} .$$

f. Current flow in conductors

In conductors made of a *homogeneous* material, σ is independent of position and we have

$$\boldsymbol{E} = -\operatorname{grad}\varphi , \qquad \boldsymbol{i} = \sigma\boldsymbol{E} = -\sigma\operatorname{grad}\varphi ,$$

$$\operatorname{div}\boldsymbol{i} = 0 , \qquad \operatorname{div}\boldsymbol{E} = 0 , \qquad \nabla^2\varphi = 0 ,$$

and thus $\varrho = 0$. In general, however, surface charges are present. Furthermore,

$$i_n = -\sigma \frac{\partial \varphi}{\partial n} = 0$$

at the surface.

In conductors made of *inhomogeneous* materials one has

$$\operatorname{div}(\sigma \operatorname{grad} \varphi) = 0 \qquad (\varrho \neq 0)\,. \tag{14.11}$$

This equation is very similar to that for fluid flow. The lines of current flow are, therefore, essentially the streamlines of hydrodynamic potential flow.

Mathematically, the problem of current distribution is as follows. The potential difference between the electrodes,

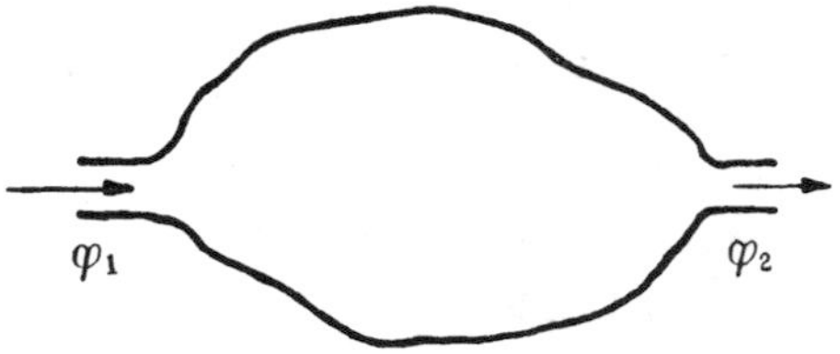

Figure 14.2

$\varphi_1 - \varphi_2$, is given. At the surfaces, $\sigma(\partial\varphi/\partial n) = 0$. In the interior we have the differential equation [14.11]. These requirements determine the current distribution [A-3]. Since $\sigma(\partial\varphi/\partial n)$ is nonvanishing only at electrodes 1 and 2, it follows from Gauss's law that

$$+\int_1 \sigma \frac{\partial \varphi}{\partial n}\, \mathrm{d}f = -\int_2 \sigma \frac{\partial \varphi}{\partial n}\, \mathrm{d}f = J\,.$$

For fixed geometry, if φ is multiplied by a constant factor then the current also changes by the same factor. From this it is seen that for arbitrary conductors,

$$\varphi_1 - \varphi_2 = JR\,.$$

We can now also prove Joule's law in a more general manner. The total heat developed in a conductor is

$$\bar{Q} = \int \sigma E^2 \,\mathrm{d}V = \int \boldsymbol{i}\cdot\boldsymbol{E}\,\mathrm{d}V = -\int \boldsymbol{i}\cdot\operatorname{grad}\varphi\,\mathrm{d}V$$

$$= +\int \varphi \operatorname{div}\boldsymbol{i}\,\mathrm{d}V - \int \operatorname{div}(\varphi\boldsymbol{i})\,\mathrm{d}V = -\oint \varphi i_n\,\mathrm{d}f\,.$$

Since $i_n \neq 0$ only at the electrodes, we have

$$\bar{Q} = (\varphi_1 - \varphi_2)J = J^2 R\,.$$

If several electrodes are present a sum is obtained.

15. THE MAGNETIC FIELDS OF STEADY-STATE CURRENTS

For the present we can again think of the magnetic field as being measurable with magnetic dipoles.

In his famous experiment Oersted found that the line integral of the magnetic field produced by a steady-state current, taken over a closed curve enclosing the current, is given by

$$\oint H_s\,\mathrm{d}s = \frac{4\pi}{c} J = 4\pi J_{\mathrm{emu}}\,. \qquad [15.1]$$

Here, c is the velocity of light, J the current measured in esu, and J_{emu} the current measured in emu. For an infinitely long current-carrying cylinder, the radial and azimuthal components of the magnetic field outside the cylinder are, in polar coordinates,

$$H_r = 0\,, \qquad H_\varphi = \frac{2J}{c}\frac{1}{r}\,. \qquad [15.2]$$

The relation between the sense of rotation and the direction of current flow is shown in Fig. 15.1. This association is, however, necessary only if $\boldsymbol{H}$ is considered as a vector.

On the other hand, if $\boldsymbol{H}$ is considered as an antisymmetric tensor, then the association is not necessary.

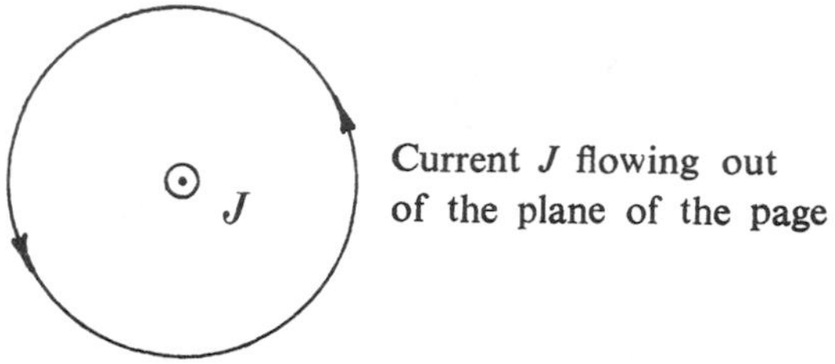

Figure 15.1

As a consequence of its introduction in analogy to the electric field intensity, the dimensions of $\boldsymbol{H}$ are

$$[H] = [E] = \text{charge}\cdot\text{cm}^{-2}\,.$$

From

$$[J] = \text{charge}\cdot\text{sec}^{-1}$$

it then follows that

$$[c] = \text{cm}\cdot\text{sec}^{-1}\,.$$

Thus, c has indeed the dimensions of velocity.

We will consider the above relations as a *fundamental law* for steady-state currents. Corresponding to the equations for $\boldsymbol{E}$,

$$\oint E_n\,\mathrm{d}f = 4\pi e\,, \qquad \operatorname{curl}\boldsymbol{E} = 0\,,$$

we now have the analogous relations for $\boldsymbol{H}$:

$$\oint H_s\,\mathrm{d}s = \frac{4\pi}{c}\,J\,, \qquad \operatorname{div}\boldsymbol{H} = 0\,. \qquad [15.3]$$

(The second equation for $\boldsymbol{H}$ requires that $H_r = 0$, since $\oint_{\text{cylinder}} H_n\,\mathrm{d}f = 2\pi r H_r = 0$.) Equation [15.1] is valid within a conductor as well as without:

$$\oint_C H_s\,\mathrm{d}s = \frac{4\pi}{c}\,J_F = \frac{4\pi}{c}\int_F i_n\,\mathrm{d}f\,, \qquad [15.4]$$

where J_F is the current flowing through the surface F enclosed by the arbitrary curve C. If the conductor is *cylindrically symmetric* with radius a, then we know that in the steady-

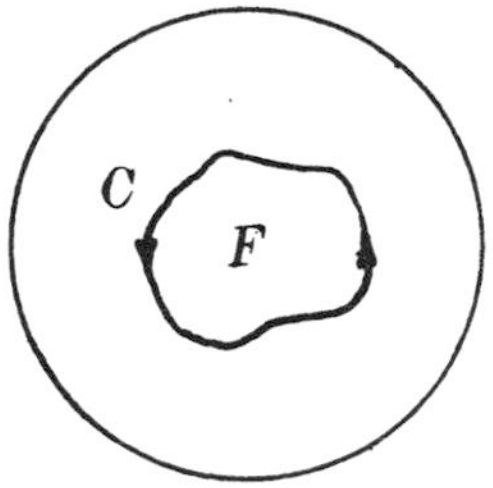

Figure 15.2

state the current density is uniform [A-3]. If the curve C is a circle of radius r, then

$$J_F = J\left(\frac{r}{a}\right)^2 \qquad (r < a)\,,$$

$$2\pi r H_\varphi = \frac{4\pi}{c} J\left(\frac{r}{a}\right)^2 : \quad \begin{cases} H_\varphi = \dfrac{2J}{c}\dfrac{r}{a^2} & (r \leqslant a)\,, \\[2ex] H_\varphi = \dfrac{2J}{c}\dfrac{1}{r} & (r \geqslant a)\,. \end{cases}$$

Application: the solenoid. We will assume that the spacing between individual turns is small compared with both the length h and the diameter of the solenoid. For this limiting case we can roughly approximate the real field by one that is uniform inside the solenoid and vanishes outside. For the path of integration shown in Fig. 15.3,

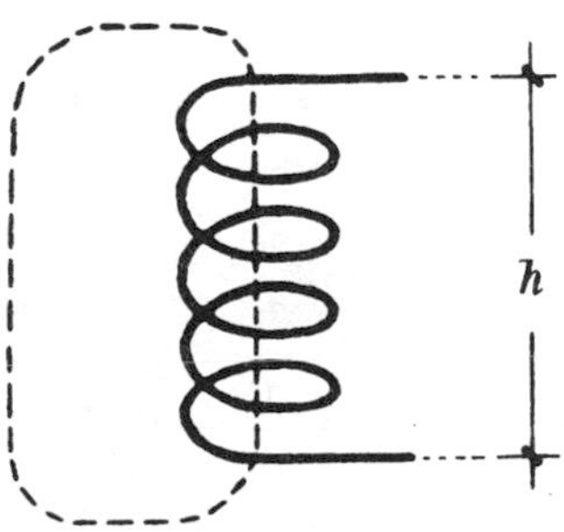

Figure 15.3

then,

$$\int i_n \,\mathrm{d}f = NJ; \qquad \oint H_s \,\mathrm{d}s = Hh = \frac{4\pi}{c} NJ ,$$

where N is the number of turns and J the current per turn. Thus,

$$H = \frac{4\pi N}{h} \frac{J}{c} . \tag{15.5}$$

The magnetic field within the solenoid is directly proportional to the current.

Just as to the equation $\oint E_n \mathrm{d}f = 4\pi e$ there corresponds the equivalent requirement $\mathrm{div}\boldsymbol{E} = 4\pi\varrho$, there exists a differential form corresponding to the integral relation $\oint H_s \mathrm{d}s = 4\pi J_F/c$ of Eq. [15.4]. According to Stokes's theorem,

$$\int \mathrm{curl}_n \boldsymbol{H} \,\mathrm{d}f = \frac{4\pi}{c} \int i_n \,\mathrm{d}f .$$

Since this is valid for arbitrary surfaces, then

$$\mathrm{curl}\, \boldsymbol{H} = \frac{4\pi}{c} \boldsymbol{i} = 4\pi \boldsymbol{i}_{\mathrm{emu}} . \tag{15.6}$$

This is the differential equation which determines the magnetic field for a given steady-state current.

Important Remark. We have seen that $\mathrm{div}\,\mathrm{curl}\,\boldsymbol{a} = 0$ for any arbitrary vector $\boldsymbol{a}$. Thus, Eq. [15.6] is correct only as long as $\mathrm{div}\,\boldsymbol{i} = 0$. This, however, is no longer true for the time-dependent case since, in general, $\partial\varrho/\partial t \neq 0$. Hence, the relation between $\boldsymbol{i}$ and $\boldsymbol{H}$ as expressed in the form of Eq. [15.6] cannot be correct in this case.

Calculation of $\boldsymbol{H}$ *with the help of the vector potential* $\boldsymbol{A}$

The requirement $\mathrm{div}\boldsymbol{H} = 0$ is automatically satisfied by setting

$$\boldsymbol{H} = \mathrm{curl}\, \boldsymbol{A} . \tag{15.7}$$

The vector potential $\boldsymbol{A}$ must now be determined. As already mentioned in Section 12, $\boldsymbol{A}$ is arbitrary to within the gradient of a scalar function. That is, for

$$\boldsymbol{A}' = \boldsymbol{A} + \operatorname{grad} \psi$$

with ψ arbitrary, $\boldsymbol{H} = \operatorname{curl} \boldsymbol{A}'$ is also true. The function ψ can be determined by requiring, for example, that

$$\operatorname{div} \boldsymbol{A} = 0 . \tag{15.8}$$

This is convenient for the case of steady-state currents.

Because of the vector identity

$$\operatorname{curl} \operatorname{curl} \boldsymbol{a} \equiv \operatorname{grad} \operatorname{div} \boldsymbol{a} - \nabla^2 \boldsymbol{a} , \tag{15.9}$$

the vector potential must satisfy the condition

$$\operatorname{grad} \operatorname{div} \boldsymbol{A} - \nabla^2 \boldsymbol{A} = \frac{4\pi}{c} \boldsymbol{i} . \tag{15.10}$$

(As employed here, the symbol ∇^2 represents the Laplacian operator expressed in cartesian coordinates and applied to each component of the vector $\boldsymbol{a}$. In curvilinear coordinates the situation is somewhat more complicated.) Equation [15.10] is satisfied if the following conditions are fulfilled:

$$\operatorname{div} \boldsymbol{A} = 0, \qquad \nabla^2 \boldsymbol{A} = -\frac{4\pi}{c} \boldsymbol{i} . \tag{15.11}$$

Each component of the second part of Eq. [15.11] is analogous to the expression $\nabla^2\varphi = -4\pi\varrho$ of Eq. [3.5] which we know how to integrate: $\varphi_P = \int \varrho_Q \, dV_Q / r_{PQ}$. Therefore,

$$\boldsymbol{A}_P = \frac{1}{c} \int \frac{\boldsymbol{i}_Q \, dV_Q}{r_{PQ}} . \tag{15.12}$$

We have still to show that $\operatorname{div} \boldsymbol{A} = 0$. By twice applying the identity [4.10],

$$\operatorname{div}(f\boldsymbol{a}) \equiv f \operatorname{div} \boldsymbol{a} + \boldsymbol{a} \cdot \operatorname{grad} f ,$$

we obtain

$$\operatorname{div}_P\left(\frac{\boldsymbol{i}_Q}{r_{PQ}}\right)=\left(\operatorname{grad}_P\frac{1}{r_{PQ}}\right)\cdot\boldsymbol{i}_Q=-\left(\operatorname{grad}_Q\frac{1}{r_{PQ}}\right)\cdot\boldsymbol{i}_Q$$

$$=-\operatorname{div}_Q\left(\frac{\boldsymbol{i}_Q}{r_{PQ}}\right)+\frac{1}{r_{PQ}}\operatorname{div}_Q\boldsymbol{i}\,.$$

Thus,

$$\operatorname{div}_P\boldsymbol{A}=-\frac{1}{c}\int\operatorname{div}_Q\left(\frac{\boldsymbol{i}_Q}{r_{PQ}}\right)\mathrm{d}V_Q+\frac{1}{c}\int\frac{1}{r_{PQ}}(\operatorname{div}\boldsymbol{i})_Q\,\mathrm{d}V_Q$$

$$=-\frac{1}{c}\oint\frac{i_n}{r_{PQ}}\,\mathrm{d}f_Q+\frac{1}{c}\int\frac{1}{r_{PQ}}(\operatorname{div}\boldsymbol{i})_Q\,\mathrm{d}V_Q=0\,,$$

since $i_n=0$ and $\operatorname{div}\boldsymbol{i}=0$.

Now, in order to calculate $\boldsymbol{H}$ from $\boldsymbol{A}$ we make use of the identity

$$\operatorname{curl}(f\boldsymbol{a})\equiv f\operatorname{curl}\boldsymbol{a}+\operatorname{grad}f\times\boldsymbol{a}\,. \qquad [15.13]$$

Then,

$$\boldsymbol{H}_P=\operatorname{curl}_P\boldsymbol{A}=\frac{1}{c}\int\operatorname{grad}_P\frac{1}{r_{PQ}}\times\boldsymbol{i}_Q\,\mathrm{d}V_Q\,,$$

since curl_P does not operate on $\boldsymbol{i}_Q$. Hence,

$$\boldsymbol{H}_P=\frac{1}{c}\int\boldsymbol{i}_Q\times\frac{(\boldsymbol{x}_P-\boldsymbol{x}_Q)}{r_{PQ}^3}\,\mathrm{d}V_Q=\frac{1}{c}\int\boldsymbol{i}_Q\times\frac{\boldsymbol{t}}{r_{PQ}^2}\,\mathrm{d}V_Q\,, \qquad [15.14]$$

in which $\boldsymbol{t}$ represents the unit vector $\boldsymbol{t}=(\boldsymbol{x}_P-\boldsymbol{x}_Q)/r_{PQ}$.

Limiting case: a linear conductor

For a conductor whose cross-sectional dimensions are very small compared with its length, the following mathe-

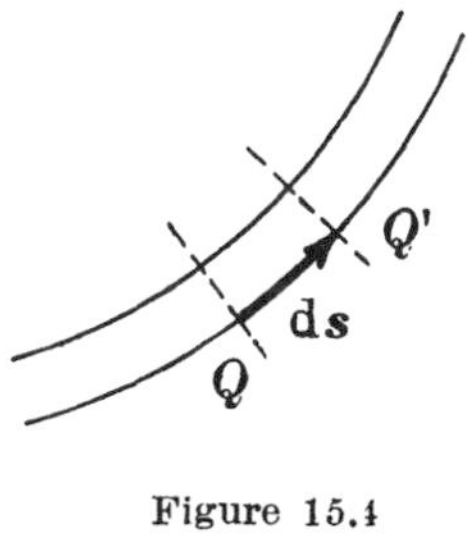

Figure 15.4

matical idealization (which, however, is not always admissible) can be carried out.

Let Q be the cross-sectional area of the conductor and $\mathbf{ds}$ the line element (see Fig. 15.4). Then,

$$\int\limits_{\text{(conductor)}} \boldsymbol{i}\,\mathrm{d}V = \int\limits_{\text{(conductor)}} \boldsymbol{i}\,\mathrm{d}f\,|\mathrm{d}\boldsymbol{s}| = \int\limits_{s}\int\limits_{Q} i_n\,\mathrm{d}f\,\mathrm{d}\boldsymbol{s} = \int\limits_{s} J\,\mathrm{d}\boldsymbol{s}\,,$$

since $\boldsymbol{i}$ is parallel to $\mathrm{d}\boldsymbol{s}$. Then,

$$\boldsymbol{A}_P = \frac{J}{c}\int \frac{\mathrm{d}\boldsymbol{s}_Q}{r_{PQ}} \qquad [15.15]$$

and

$$\boldsymbol{H}_P = \frac{J}{c}\int \frac{\mathrm{d}\boldsymbol{s}_Q \times (\boldsymbol{x}_P - \boldsymbol{x}_Q)}{r_{PQ}^3} = \frac{J}{c}\int \mathrm{d}\boldsymbol{s}_Q \times \frac{\boldsymbol{t}}{r_{PQ}^2}\,. \qquad [15.16]$$

This expression is familiar under the name of the ***Biot-Savart law.***

It is often said that the magnetic field produced by the current element $J\,\mathrm{d}\boldsymbol{s}$ is

$$\text{“}\mathrm{d}\boldsymbol{H}\text{”} = \frac{J}{c}\,\mathrm{d}\boldsymbol{s} \times \frac{\boldsymbol{t}}{r^2}\,.$$

However, the “$\mathrm{d}\boldsymbol{H}$” so defined does not satisfy the equations from which we started, since the current element does not satisfy the continuity equation. In reality there are no isolated current elements. Consequently, ***the Biot-Savart law***

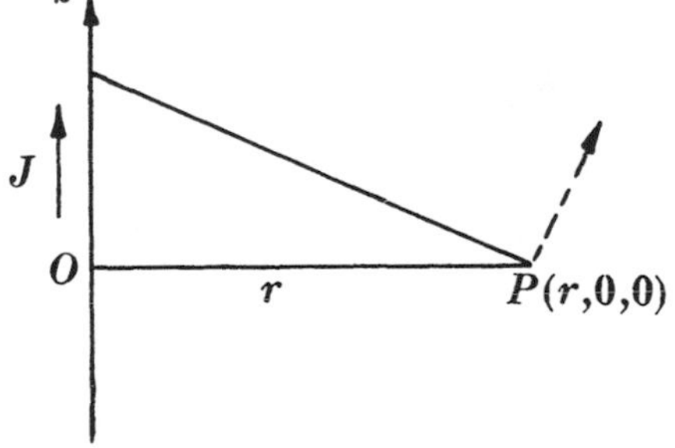

Figure 15.5

can be applied only to closed current loops or to infinitely long conductors, the latter of which can be thought of as being closed at infinity.

As a test, we apply the Biot-Savart law to an infinitely long linear conductor (Fig. 15.5). If the coordinates are chosen so that $d\boldsymbol{s}$ is parallel to the z axis, then $x_P - x_Q = r =$ constant and

$$H_y = \frac{J}{c}\, r \int_{-\infty}^{+\infty} \frac{dz}{(z^2+r^2)^{\frac{3}{2}}} = \frac{J}{c}\, r \cdot \left. \frac{z\dfrac{1}{r^2}}{(r^2+z^2)^{\frac{1}{2}}} \right|_{-\infty}^{+\infty} = 2\frac{J}{c}\frac{1}{r}.$$

This is the formula which we have already obtained in Eq. [15.2].

16. THE EQUIVALENCE OF THE MAGNETIC FIELD DUE TO CLOSED CURRENT LOOPS WITH THAT DUE TO A DISTRIBUTION OF MAGNETIC DIPOLES

We will first consider a transformation of the *vector potential due to a closed linear current loop*, which leads to the concept of magnetic moment. For this we require a special form of Stokes's theorem. According to Eq. [15.13], for a vector $\boldsymbol{a}$ which is independent of position,

$$\operatorname{curl}(f\boldsymbol{a}) = \operatorname{grad} f \times \boldsymbol{a}\,.$$

Applying Stokes's theorem (Eq. [2.5]) to the vector $f\boldsymbol{a}$ (where $\boldsymbol{a}$ is constant), we get

$$\oint (f\boldsymbol{a})\cdot d\boldsymbol{s} = \int (\operatorname{grad} f \times \boldsymbol{a})\cdot \boldsymbol{n}\, df\,,$$

$$\boldsymbol{a}\cdot\oint f\, d\boldsymbol{s} = \boldsymbol{a}\cdot\int \boldsymbol{n}\times \operatorname{grad} f\, df\,.$$

Since $\boldsymbol{a}$ is an arbitrary vector, it must be true that

$$\oint f\, d\boldsymbol{s} = \int \boldsymbol{n}\times \operatorname{grad} f\, df \qquad [16.1]$$

For the vector potential $\boldsymbol{A}_P = (J/c)\oint \mathrm{d}\boldsymbol{s}_Q/r_{PQ}$, this integral relation results in

$$\boldsymbol{A}_P = \frac{J}{c}\int \boldsymbol{n}\times \mathrm{grad}_Q \frac{1}{r_{PQ}}\,\mathrm{d}f\,. \qquad [16.2]$$

Here, only the bounding curve (i.e., the conducting loop) of the surface over which the integration is performed is determined; the surface itself is arbitrary. On the other hand, we have seen in Section 12 that the vector potential due to a dipole of moment $\boldsymbol{m}$ is

$$\boldsymbol{A}_P = \boldsymbol{m}\times \mathrm{grad}_Q \frac{1}{r_{PQ}}\,. \qquad [16.3]$$

Thus, it is seen that *the vector potential, and therefore the magnetic field due to a closed linear current loop, is equivalent to a surface distribution of dipoles with a magnetic moment per unit area of*

$$\boldsymbol{m} = \frac{J}{c}\boldsymbol{n}\,,$$

where $\boldsymbol{n}$ is the normal to the surface. The choice of the surface is arbitrary, except that the bounding curve is determined by the current loop.

Special case: a plane current loop. As the surface, we choose the plane area enclosed by the loop. On this surface the distribution of magnetic moment is uniform. A further simplification results if the field point P is located very far from the current: $r_{PQ} \gg a$, where a is, for example, the diameter of the loop. Thus, for all points Q on the surface, r_{PQ} can be replaced by r_{PO}, where O can be any fixed point on the surface. Then, if F is the area of the enclosed surface,

$$\boldsymbol{A}_P = -\frac{J}{c}F\boldsymbol{n}\times \mathrm{grad}_P \frac{1}{r_{PO}} = -\boldsymbol{M}\times \mathrm{grad}_P \frac{1}{r_{PO}}\,,$$

$$\boldsymbol{M} = \frac{J}{c}F\boldsymbol{n}\,.$$

Thus, the magnetic field of a closed current loop is, at large distances, equivalent to that of a dipole of magnetic moment $\boldsymbol{M}$ as given by the above equation.

What we have just done for linear current loops can also be done analogously for volume distributions of current, as we shall soon see. These equivalences were discovered by Ampère. It was also Ampère's idea that in reality it is not magnetic dipoles but, rather, current loops which produce magnetic fields. In this theory the elementary quantity is not the magnetic dipole; instead, magnetism is attributed to microscopic currents (called *Amperean currents*) within molecules and atoms.

We now consider a ***volume distribution of dipoles*** by a generalization of the preceding considerations. If the magnetic moment per unit volume (the magnetization) is denoted by $\boldsymbol{M}$, then the vector potential of the field produced is

$$\boldsymbol{A} = \int \boldsymbol{M} \times \operatorname{grad}_Q \frac{1}{r} \, \mathrm{d}V . \tag{16.4}$$

We wish to transform this expression to

$$\boldsymbol{A} = \frac{1}{c} \int \frac{\boldsymbol{i}_m}{r} \, \mathrm{d}V + \frac{1}{c} \int \frac{\boldsymbol{j}_m}{r} \, \mathrm{d}f . \tag{16.5}$$

This represents a vector potential produced by molecular volume currents $\boldsymbol{i}_m$ and molecular surface currents $\boldsymbol{j}_m$. In order to carry out this transformation we need an integral relation which follows from Gauss's theorem. Because of the identity

$$\operatorname{div}(\boldsymbol{a} \times \boldsymbol{b}) \equiv \boldsymbol{b} \operatorname{curl} \boldsymbol{a} - \boldsymbol{a} \operatorname{curl} \boldsymbol{b} , \tag{16.6}$$

Gauss's theorem of Eq. [2.9], with the help of a constant vector $\boldsymbol{a}$, yields

$$\boldsymbol{a} \cdot \oint \boldsymbol{b} \times \boldsymbol{n} \, \mathrm{d}f = \oint (\boldsymbol{a} \times \boldsymbol{b}) \cdot \boldsymbol{n} \, \mathrm{d}f = -\boldsymbol{a} \cdot \int \operatorname{curl} \boldsymbol{b} \, \mathrm{d}V .$$

Since $\boldsymbol{a}$ is arbitrary,

$$\oint \boldsymbol{n} \times \boldsymbol{b} \, \mathrm{d}f \equiv \int \operatorname{curl} \boldsymbol{b} \, \mathrm{d}V . \qquad [16.7]$$

With the aid of this integral expression and the fact that

$$\operatorname{curl}(f\boldsymbol{a}) = f \operatorname{curl} \boldsymbol{a} + \operatorname{grad} f \times \boldsymbol{a} ,$$

we obtain from Eq. [16.4]

$$\begin{aligned} \boldsymbol{A} &= \int \frac{1}{r} \operatorname{curl} \boldsymbol{M} \, \mathrm{d}V - \int \operatorname{curl}\left(\frac{\boldsymbol{M}}{r}\right) \mathrm{d}V \\ &= \int \frac{1}{r} \operatorname{curl} \boldsymbol{M} \, \mathrm{d}V - \oint \frac{\boldsymbol{n} \times \boldsymbol{M}}{r} \, \mathrm{d}f . \end{aligned}$$

A volume distribution of magnetization is, therefore, equivalent to a volume current density $\boldsymbol{i}_m$ *and a surface current density* $\boldsymbol{j}_m$, *where*

$$\boldsymbol{i}_m = c \operatorname{curl} \boldsymbol{M} , \qquad [16.8]$$

$$\boldsymbol{j}_m = -c \, \boldsymbol{n} \times \boldsymbol{M} . \qquad [16.9]$$

The previously considered case of the linear current loop is contained in this result as a special case. ($\boldsymbol{M}$ constant in the interior; in passing to the two-dimensional case the surface current goes over into a linear current, and there is a magnetic moment per unit area rather than per unit volume.)

The equivalence of the molecular currents $\boldsymbol{i}_m$ and $\boldsymbol{j}_m$ with a distribution of magnetic moment indicates the possibility of attributing magnetism to such molecular currents, as was done by Ampère.

To differentiate it from these *molecular currents*, we shall designate the quantity previously referred to simply as "current" by the name *conduction current* and write the corresponding current density as $\boldsymbol{i}_c$. In addition, there is yet a third kind of current, the so-called *convection current*, which originates when a macroscopic charge is in motion. If ϱ is the volume charge density of the moving body and $\boldsymbol{v}$

its velocity, then the convection current density is given by

$$\boldsymbol{i}_k = \varrho \boldsymbol{v} . \qquad [16.10]$$

It was shown by Rowland that the magnetic field produced by such a convection current is identical with that produced by an ordinary conduction current; that is, we have

$$\operatorname{curl} \boldsymbol{H} = \frac{4\pi}{c} (\boldsymbol{i}_c + \boldsymbol{i}_k) . \qquad [16.11]$$

The equivalence of molecular currents with a distribution of magnetic dipoles as discussed in these paragraphs now demonstrates the second analogy to dielectrics mentioned in Section 12. That is, *the average value of the microscopic magnetic field intensity must be identified with the magnetic induction* $\boldsymbol{B}$, *and not with* $\boldsymbol{H}$. This follows from the fact that for the steady-state case,

$$(1) \qquad \operatorname{div} \boldsymbol{B} = 0 ,$$

$$(2) \qquad \operatorname{curl} \boldsymbol{B} = \frac{4\pi}{c} \boldsymbol{i} = \frac{4\pi}{c} (\boldsymbol{i}_c + \boldsymbol{i}_k + \boldsymbol{i}_m) . \qquad [16.12]$$

Equation [16.12] follows from the relation $\boldsymbol{B} = \boldsymbol{H} + 4\pi \boldsymbol{M}$ of [12.2], and from Eqs. [16.11] and [16.8] for curl $\boldsymbol{H}$ and curl $\boldsymbol{M}$.

The induction $\boldsymbol{B}$ has discontinuous tangential components if surface currents $\boldsymbol{j}_m$ are present. By applying Stokes's law to the integration path shown in Fig. 16.1, it is found

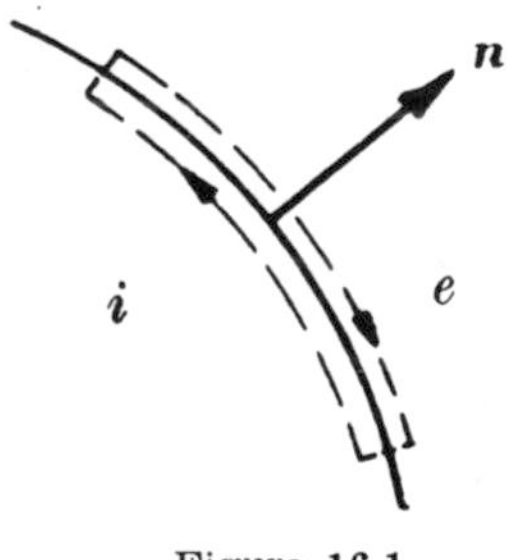

Figure 16.1

that

$$\boldsymbol{B}_{\parallel}^{(i)}-\boldsymbol{B}_{\parallel}^{(e)}=\frac{4\pi}{c}\,\boldsymbol{n}\times\boldsymbol{j}_m=4\pi\boldsymbol{M}_{\parallel}\,. \qquad [16.13]$$

Here $\boldsymbol{n}$ is a unit vector in the normal direction and $\boldsymbol{M}_{\parallel}$ is the tangential component of $\boldsymbol{M}$ at the surface.

The vortices (curls) of the $\boldsymbol{B}$ field are thus given by *all* the currents present. Analogously, the sources (divergence) of $\boldsymbol{E}$ were determined by *all* the charges present, and $\boldsymbol{E}$ was the average of the microscopic electric field intensity. Therefore, in this deeper analogy $\boldsymbol{B}$ corresponds to $\boldsymbol{E}$. However, $\boldsymbol{B}$ is a source-free vortex field while $\boldsymbol{E}$ (in the quasi-static case) is a vortex-free source field. (The operators curl and div are, therefore, also interchanged.)

In analogy to $\boldsymbol{D}=\boldsymbol{E}+4\pi\boldsymbol{P}$, the equation $\boldsymbol{H}=\boldsymbol{B}-4\pi\boldsymbol{M}$ is, from this Amperean viewpoint, only an auxiliary relation. At a bounding surface the normal component of $\boldsymbol{B}$ is continuous while $\boldsymbol{H}$ has a continuous tangential component. The more fundamental analogy between magnetic and electric quantities is, therefore,

$\boldsymbol{B}$	$\boldsymbol{E}$
$\boldsymbol{H}$	$\boldsymbol{D}$
$-\boldsymbol{M}$	$\boldsymbol{P}$.

17. PONDEROMOTIVE FORCES

Until now we have defined $\boldsymbol{H}$ as being the force acting on a unit magnetic pole. However, $\boldsymbol{H}$ can also be defined in terms of the force exerted on a current-carrying con-

Figure 17.1

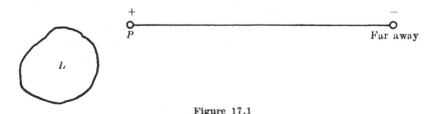

ductor element. The forces exerted by magnetic fields on conductors are called *magnetic ponderomotive forces.*

The force acting on a unit pole at P (Fig. 17.1) due to the conductor L is

$$\boldsymbol{K}_P = \frac{1}{c}\int \frac{\boldsymbol{i}_Q \times (\boldsymbol{x}_P - \boldsymbol{x}_Q)}{r_{PQ}^3}\,\mathrm{d}V_Q\,. \qquad [17.1]$$

On the other hand, the magnetic field produced by the unit pole at a point Q on the conductor is

$$\boldsymbol{H}_{QP} = \frac{\boldsymbol{x}_Q - \boldsymbol{x}_P}{r_{PQ}^3}\,.$$

According to the *principle of action and reaction* (action = reaction), the force on the conductor due to the unit pole at P must satisfy

$$\boldsymbol{K}_L = \int \boldsymbol{k}_{QP}\,\mathrm{d}V_Q = -\,\boldsymbol{K}_P\,,$$

where $\boldsymbol{K}_L$ represents the total force (i.e., integrated over all points Q). Thus,

$$\boldsymbol{K}_L = \frac{1}{c}\int \boldsymbol{i}_Q \times \boldsymbol{H}_{QP}\,\mathrm{d}V_Q\,.$$

It now appears that the force exerted by the magnetic field $\boldsymbol{H}_{QP}$ on the volume element $\mathrm{d}V_Q$ is given by $(1/c)$ $\cdot(\boldsymbol{i}_Q \times \boldsymbol{H}_{QP})\mathrm{d}V_Q$. This consideration is, however, only heuristic. In the first place, the assumption of a unit magnetic pole is not very satisfying; nevertheless, it can be realized if the second pole is sufficiently far away. The chief reason, however, why the above consideration cannot be rigorous is that the principle of action and reaction is applicable only to the whole conductor, and the force on an individual volume element cannot be obtained from it.

It turns out, however, that the *force per unit volume* is indeed given by

$$\boldsymbol{k} = \frac{1}{c}\,\boldsymbol{i} \times \boldsymbol{H}\,. \qquad [17.2]$$

This is a new *fundamental law of nature.* From this it follows that the ***magnetic force on a linear conducting element*** $\mathrm{d}\boldsymbol{s}$ is

$$\mathrm{d}\boldsymbol{K} = \frac{J}{c}\,\mathrm{d}\boldsymbol{s} \times \boldsymbol{H}. \qquad [17.3]$$

From this, in turn, it can be shown that the torque experienced by a linear conducting loop carrying a current J and located in a uniform magnetic field $\boldsymbol{H}$ is the same as that acting on a rigid magnetic dipole of moment $\boldsymbol{M} = (J/c)\int_F \boldsymbol{n}\,\mathrm{d}f$; that is $\boldsymbol{D} = \boldsymbol{M} \times \boldsymbol{H}$. Ampère performed careful investigations on the magnetic ponderomotive forces acting on conductors and thereby demonstrated the complete equivalence—also with respect to the effects of these forces—of closed current loops with dipole distributions.

The above force law is also valid for convection currents $\boldsymbol{i}_k = \varrho \boldsymbol{v}$ (where ϱ is the volume charge density of the moving body and $\boldsymbol{v}$ its velocity). Thus, the total force $\boldsymbol{K} = \int \boldsymbol{k}\,\mathrm{d}V$ acting on a body of charge $e = \int \varrho\,\mathrm{d}V$ is

$$\boldsymbol{K} = \frac{e}{c}\,\boldsymbol{v} \times \boldsymbol{H}. \qquad [17.4]$$

This is called the *Lorentz force* after the Dutch physicist H. A. Lorentz. This relation shows that $\boldsymbol{H}$ ***can be defined and measured with the aid of moving charges***, just as $\boldsymbol{E}$ can be defined and measured with stationary charges.

The following force law is thus valid for the general case of an ***electric and a magnetic field***: the force per cm³ is composed of

$$\boldsymbol{k}_e = \varrho \boldsymbol{E} \qquad \text{(the electric force)}$$

and

$$\boldsymbol{k}_m = \frac{1}{c}\,\boldsymbol{i} \times \boldsymbol{H} \quad \text{(the magnetic force)}. \qquad [17.5]$$

Here, $\boldsymbol{i}$ is the total current density, including convection currents if present. This law is quite general and is not restricted to steady-state currents.

The *total force acting on a charge e* consists of

$$\boldsymbol{K}_e = e\boldsymbol{E}$$

and

$$\boldsymbol{K}_m = \frac{e}{c}\,\boldsymbol{v}\times\boldsymbol{H}\,. \qquad [17.6]$$

The validity of this force law has been confirmed by the deflection of cathode rays (electrons) by electric and magnetic fields. It should also be remarked that the Lorentz force does no work since it is directed normal to the velocity.

If $\boldsymbol{H}_1$ is produced by a closed linear current loop (denoted by 2), then

$$\boldsymbol{H}_1 = \frac{J_2}{c}\oint_2 \frac{\mathrm{d}\boldsymbol{s}_2\times(\boldsymbol{x}_1-\boldsymbol{x}_2)}{r_{12}^3}\,, \qquad [17.7]$$

and the magnetic force exerted on a linear current element $\mathrm{d}\boldsymbol{s}_1$ of a second conductor (denoted by 1) is

$$\mathrm{d}\boldsymbol{K}_m^{(1)} = \frac{J_1 J_2}{c^2}\oint_2 \frac{\mathrm{d}\boldsymbol{s}_1\times\left(\mathrm{d}\boldsymbol{s}_2\times(\boldsymbol{x}_1-\boldsymbol{x}_2)\right)}{r_{12}^3}\,. \qquad [17.8]$$

Empirically, this is all that can be said (the current element $\mathrm{d}\boldsymbol{s}_1$ can be made movable). On the other hand, it is dangerous to omit the integral and to seek a fundamental law expressing the effect of one current element upon another. Before the appearance of the Maxwell theory, this was indeed done from the "action at a distance" point of view. In doing this, however, one goes beyond what can be experimentally determined, since the contribution of the other conducting elements cannot be neglected. Accordingly, the result is not unique. The only thing that can be required of such a fundamental law is that upon integration, Eq. [17.8] must follow. For example, Grassman has found the "fundamental law"

$$\mathrm{d}\boldsymbol{K}_m^{(1,2)} = \frac{J_1 J_2}{c^2}\,\frac{\mathrm{d}\boldsymbol{s}_1\times(\mathrm{d}\boldsymbol{s}_2\times\boldsymbol{n}_{21})}{r_{12}^2}\,, \qquad \text{where} \qquad \boldsymbol{n}_{21} = \frac{\boldsymbol{x}_1-\boldsymbol{x}_2}{r_{12}}\,.$$

Equation [17.8] results upon integration over $\mathrm{d}\boldsymbol{s}_2$. However, it turns out that

$$\mathrm{d}\boldsymbol{K}_m^{(1,2)} \neq -\,\mathrm{d}\boldsymbol{K}_m^{(2,1)}\,.$$

Nevertheless, the principle of action and reaction is valid for the whole conductor. There are still other "fundamental laws" which yield the same results.

Historically, the oldest form is that of Ampère which is constructed on the basis of certain axioms: $\mathrm{d}\boldsymbol{K}^{(1,2)}$ is to have the direction of $\boldsymbol{n}_{12}=(\boldsymbol{x}_2-\boldsymbol{x}_1)/r_{12}$, is to be bilinear in $\mathrm{d}\boldsymbol{s}_1$ and $\mathrm{d}\boldsymbol{s}_2$, and is to be homogeneous of degree zero in r_{12}, $\mathrm{d}\boldsymbol{s}_1$, and $\mathrm{d}\boldsymbol{s}_2$. From this it follows that it must have the form

$$\mathrm{d}\boldsymbol{K}^{(1,2)} = \frac{J_1 J_2}{c^2}\frac{\boldsymbol{n}_{12}}{r_{12}^2}\{\alpha(\mathrm{d}\boldsymbol{s}_1\cdot\mathrm{d}\boldsymbol{s}_2) + \beta(\mathrm{d}\boldsymbol{s}_1\cdot\boldsymbol{n}_{12})(\mathrm{d}\boldsymbol{s}_2\cdot\boldsymbol{n}_{12})\}\,.$$

The coefficients α and β are determined from the requirement that

$$\oint_2 \mathrm{d}\boldsymbol{K}^{(1,2)} \equiv \mathrm{d}\boldsymbol{K}_m^{(1)}\,,$$

in accordance with Eq. [17.8].

These "fundamental laws" completely lack physical sig nificance, since in the case of time-dependent currents it is not at all easy to say from which current element the field (which is, indeed, a measure of the force) originates. From the point of view of field theory, one cannot inquire about such "fundamental laws."

18. THE PRINCIPLE OF ACTION AND REACTION FOR ELECTRIC AND MAGNETIC FORCES. THE MAXWELL STRESS TENSORS

In this section, only volume distributions of current and charge are considered.

We first introduce the concept of a *tensor*. By a tensor

in a 3-dimensional space is meant a system of 9 quantities $T_{ik}(i, k = 1, 2, 3)$ which, under a transformation to another coordinate system, transform as the products $a_i b_k$ formed from the components of two vectors $\boldsymbol{a}$ and $\boldsymbol{b}$. (It is, however, not necessary that the T_{ik} be representable in the form $a_i b_k$.) A tensor can have special symmetry properties: if

$T_{ik} = T_{ki}$, the tensor is symmetric;

$S_{ik} = -S_{ki}$, the tensor is skew- or antisymmetric.

An antisymmetric tensor in a 3-dimensional space transforms like a vector if only rotations (but not inversions) of the coordinate system are considered.[3] Since S_{ik} transforms as $a_i b_k - a_k b_i$, an antisymmetric tensor transforms as a vector product $\boldsymbol{a} \times \boldsymbol{b}$.

Force is an ordinary (polar) vector, as is the current density $\boldsymbol{i}$. Thus, according to Eq. [17.2], $\boldsymbol{H}$ is obviously an antisymmetric tensor. This can also be seen from the Biot-Savart law of Eq. [15.16]. Actually, one should write H_{23}, H_{31}, and H_{12} for H_x, H_y, and H_z, respectively (with $H_{ik} = -H_{ki}$). On the other hand, the electric field intensity is an ordinary vector.

We denote the scalar product of the tensor T with the unit vector $\boldsymbol{n}$ by the vector

$$\boldsymbol{T}_n \equiv (T_{n1}, T_{n2}, T_{n3}),$$

in which

$$T_{ni} = \sum_k T_{ik} n_k \qquad (i = 1, 2, 3).$$

It is immediately seen that the T_{ni} transform as the components of a vector.

Gauss's theorem has the form

$$\int \sum_k \frac{\partial T_{ik}}{\partial x_k} \, \mathrm{d}V = \oint \sum_k T_{ik} n_k \, \mathrm{d}f, \qquad [18.1]$$

[3] See the discussion on pp. 11 f.

which can also be written as

$$\int \operatorname{div} T \, dV = \oint \boldsymbol{T}_n \, df , \tag{18.2}$$

where

$$\operatorname{div} T = \left(\sum_k \frac{\partial T_{1k}}{\partial x_k} , \sum_k \frac{\partial T_{2k}}{\partial x_k} , \sum_k \frac{\partial T_{3k}}{\partial x_k} \right) . \tag{18.3}$$

The principle of action and reaction is satisfied for electric as well as for magnetic forces if $\boldsymbol{k}_e$ and $\boldsymbol{k}_m$ can be represented by

$$\boldsymbol{k}_e = \operatorname{div} T^{(e)} , \qquad \boldsymbol{k}_m = \operatorname{div} T^{(m)} , \tag{18.4}$$

where $T^{(e)}$ and $T^{(m)}$ vanish sufficiently fast at infinity, so that

$$\int \boldsymbol{k}_e \, dV = \oint \boldsymbol{T}_n^{(e)} \, df \to 0$$

and [18.5]

$$\int \boldsymbol{k}_m \, dV = \oint \boldsymbol{T}_n^{(m)} \, df \to 0$$

as the surface bounding the region of integration becomes infinite.

We will now prove that *the principle of action and reaction is satisfied for steady-state currents and stationary charges (i.e., for time-independent fields)*. For this we employ the previously derived relations of Eqs. [17.5], [3.4], [2.4] [15.6], and [15.3]:

$$\boldsymbol{k}_e = \varrho \boldsymbol{E} = \frac{1}{4\pi} \operatorname{div} \boldsymbol{E} \cdot \boldsymbol{E} , \qquad \operatorname{curl} \boldsymbol{E} = 0 ,$$

$$\boldsymbol{k}_m = \frac{1}{c} \boldsymbol{i} \times \boldsymbol{H} = \frac{1}{4\pi} \operatorname{curl} \boldsymbol{H} \times \boldsymbol{H} , \qquad \operatorname{div} \boldsymbol{H} = 0 .$$

(We use the equations corresponding to the vacuum since we are considering the microscopic fields.) These can also be written as

$$\boldsymbol{k}_e = \frac{1}{4\pi} \{ \boldsymbol{E} \operatorname{div} \boldsymbol{E} + \operatorname{curl} \boldsymbol{E} \times \boldsymbol{E} \} ,$$

$$\boldsymbol{k}_m = \frac{1}{4\pi} \{ \boldsymbol{H} \operatorname{div} \boldsymbol{H} + \operatorname{curl} \boldsymbol{H} \times \boldsymbol{H} \} .$$

Consider the purely mathematical transformation

$$[\text{curl}\,\boldsymbol{E}\times\boldsymbol{E}]_1 = (\text{curl}_2\boldsymbol{E})E_3 - (\text{curl}_3\boldsymbol{E})E_2$$

$$= \left(\frac{\partial E_1}{\partial x_3} - \frac{\partial E_3}{\partial x_1}\right)E_3 - \left(\frac{\partial E_2}{\partial x_1} - \frac{\partial E_1}{\partial x_2}\right)E_2$$

$$= \sum_{k=1}^{3}\frac{\partial E_1}{\partial x_k}E_k - \sum_{k=1}^{3}\frac{\partial E_k}{\partial x_1}E_k\,.$$

In the last expression two terms whose sum is zero have been added. For an arbitrary index i we have

$$[\text{curl}\,\boldsymbol{E}\times\boldsymbol{E}]_i = \sum_{k=1}^{3}\left(\frac{\partial E_i}{\partial x_k} - \frac{\partial E_k}{\partial x_i}\right)E_k\,,$$

from which it follows that

$$4\pi k_{ei} = \sum_k\left\{E_i\left(\frac{\partial E_k}{\partial x_k}\right) + \left(\frac{\partial E_i}{\partial x_k} - \frac{\partial E_k}{\partial x_i}\right)E_k\right\}$$

$$= \sum_k\frac{\partial}{\partial x_k}(E_iE_k) - \frac{\partial}{\partial x_i}\frac{1}{2}E^2\,.$$

Here, E^2 is a scalar which has been obtained from the contraction of a symmetric tensor. With the help of the Kronecker delta symbol

$$\delta_{ik} = \begin{cases} 0 & \text{for} \quad i \neq k\,, \\ 1 & \text{for} \quad i = k\,, \end{cases} \qquad [18.6]$$

which is itself a symmetric tensor with respect to orthogonal transformations, the tensor

$$4\pi T_{ik}^{(e)} = E_iE_k - \delta_{ik}\tfrac{1}{2}E^2$$

can be formed. Then, it is indeed true that

$$4\pi\sum_k\frac{\partial T_{ik}^{(e)}}{\partial x_k} = \sum_k\frac{\partial}{\partial x_k}(E_iE_k) - \frac{\partial}{\partial x_i}\frac{1}{2}E^2\,.$$

Equations [18.4] are thus satisfied for

$$T_{ik}^{(e)} = \frac{1}{4\pi}\left\{E_i E_k - \frac{1}{2}\delta_{ik} E^2\right\},$$
$$T_{ik}^{(m)} = \frac{1}{4\pi}\left\{H_i H_k - \frac{1}{2}\delta_{ik} H^2\right\}. \qquad [18.7]$$

These are the ***Maxwell stress tensors.*** If all currents and charges are restricted to a finite region of space, then the fields vanish at least as fast as $1/r^2$ and the surface integrals in Eq. [18.5] of

$$\boldsymbol{T}_n^{(e)} = \frac{1}{4\pi}\left\{\boldsymbol{E}(\boldsymbol{E}\cdot\boldsymbol{n}) - \boldsymbol{n}\frac{1}{2}\boldsymbol{E}^2\right\}$$

and

$$\boldsymbol{T}_n^{(m)} = \frac{1}{4\pi}\left\{\boldsymbol{H}(\boldsymbol{H}\cdot\boldsymbol{n}) - \boldsymbol{n}\frac{1}{2}\boldsymbol{H}^2\right\}$$

approach zero as the surfaces become infinite.

The Maxwell stress tensors are *symmetric.* This implies that *the resulting torque is zero* (*conservation of angular momentum*). Proof: The torque per unit volume is $\boldsymbol{d} = \boldsymbol{x}\times\boldsymbol{k}$. Then,

$$d_{ij} = -d_{ji} = x_i k_j - x_j k_i = \sum_k \left(x_i \frac{\partial T_{jk}}{\partial x_k} - x_j \frac{\partial T_{ik}}{\partial x_k}\right)$$
$$= \sum_k \frac{\partial}{\partial x_k}(x_i T_{jk} - x_j T_{ik}) - T_{ji} + T_{ij}.$$

With

$$T_{ij} - T_{ji} = d'_{ij},$$

Gauss's theorem yields

$$\int \boldsymbol{d}\,\mathrm{d}V = \oint \boldsymbol{x}\times\boldsymbol{T}_n\,\mathrm{d}f + \int \boldsymbol{d}'\,\mathrm{d}V.$$

The first term vanishes, and if $T_{ij} = T_{ji}$, so does the second.

A scalar can be obtained from any tensor by taking the

trace:

$$T = \sum_i T_{ii} = \text{scalar}.$$

(Since T_{ik} transforms like $a_i b_k$, then T transforms like $\sum_i a_i b_i = \boldsymbol{a} \cdot \boldsymbol{b}$, which is a scalar.) In the case of the Maxwell stress tensors,

$$\sum_i T_{ii}^{(e)} = \frac{1}{4\pi}\left\{E^2 - \frac{3}{2}E^2\right\} = -\frac{1}{8\pi}E^2 = -W_e, \quad [18.8]$$

$$\sum_i T_{ii}^{(m)} = -\frac{1}{8\pi}H^2 = -W_m, \quad [18.9]$$

where W_e is the energy density of the electric field. We shall see later that W_m is the energy density of the magnetic field.

The Heaviside units are quite analogous in the electric and magnetic cases:

$$e_H = \sqrt{4\pi}e, \qquad \boldsymbol{E}_H = \frac{1}{\sqrt{4\pi}}\boldsymbol{E},$$

$$\boldsymbol{k}_e = \varrho \boldsymbol{E} = \varrho_H \boldsymbol{E}_H, \qquad \operatorname{div} \boldsymbol{E}_H = \varrho_H,$$

$$\boldsymbol{i}_H = \sqrt{4\pi}\boldsymbol{i}, \qquad \boldsymbol{H}_H = \frac{1}{\sqrt{4\pi}}\boldsymbol{H},$$

$$\boldsymbol{k}_m = \frac{1}{c}\boldsymbol{i} \times \boldsymbol{H} = \frac{1}{c}\boldsymbol{i}_H \times \boldsymbol{H}_H, \qquad \operatorname{curl} \boldsymbol{H}_H = \frac{1}{c}\boldsymbol{i}_H,$$

$$W_e = \tfrac{1}{2}E_H^2, \qquad W_m = \tfrac{1}{2}H_H^2.$$

Translator's Note: In the rationalized mks system (see note, p. 58), the equations differ by constant factors from those given above. For example, Oersted's law (Eq. [15.1]) reads

$$\oint \boldsymbol{H} \cdot \mathrm{d}\boldsymbol{s} = J,$$

while the Biot-Savart law (Eq. [15.16]) is

$$\boldsymbol{H}_P = \frac{1}{4\pi}\int \frac{\boldsymbol{i}_Q \times \boldsymbol{r}_{PQ}}{r_{PQ}^3}\,\mathrm{d}V_Q, \qquad \text{where} \qquad \boldsymbol{r}_{PQ} = \boldsymbol{x}_P - \boldsymbol{x}_Q.$$

The current is measured in amperes and the units of $\boldsymbol{H}$ are ampere-turns/m. The fields $\boldsymbol{B}$ and $\boldsymbol{H}$ are related by a constitutive equation which can be written in several ways:

$$\boldsymbol{B} = \mu \boldsymbol{H} = \varkappa_m \mu_0 \boldsymbol{H} = \mu_0(1 + \chi_m)\boldsymbol{H} = \mu_0 \boldsymbol{H} + \boldsymbol{M}.$$

Here, $\mu_0 = 4\pi \times 10^{-7}$ kg·m·coul^{-2} is the permeability of free space, μ is the permeability of the medium (in kg·m·coul^{-2}), $\varkappa_m = \mu/\mu_0$ is the relative permeability (dimensionless), $\chi_m = \varkappa_m - 1$ is the magnetic susceptibility (dimensionless), and $\boldsymbol{M}$ is the magnetization. The units of both $\boldsymbol{B}$ and $\boldsymbol{M}$ are webers/m^2, where 1 weber = 1 kg·m^2/coul·sec.

In the rationalized mks system, the following equations hold:

Electric	*Magnetic*
$\boldsymbol{k}_e = \varrho \boldsymbol{E}$,	$\boldsymbol{k}_m = \boldsymbol{i} \times \boldsymbol{H}$,
$\operatorname{div} \boldsymbol{E} = \varrho$,	$\operatorname{curl} \boldsymbol{H} = \boldsymbol{i}$,
$W_e = \frac{1}{2}\varepsilon E^2$,	$W_m = \frac{1}{2}\mu H^2$.

Chapter 3. Quasi-Static Fields

The fields to be considered in this chapter are assumed to change but little during the time required for light to traverse a distance equal to the maximum dimension of the body under consideration. Thus, the finite velocity with which the fields propagate need not be considered.

19. FARADAY'S LAW OF INDUCTION

The essential law which, in the case of quasi-static fields, must be added to the previous equations is *Faraday's law of induction.* When there are no magnetizable bodies present,

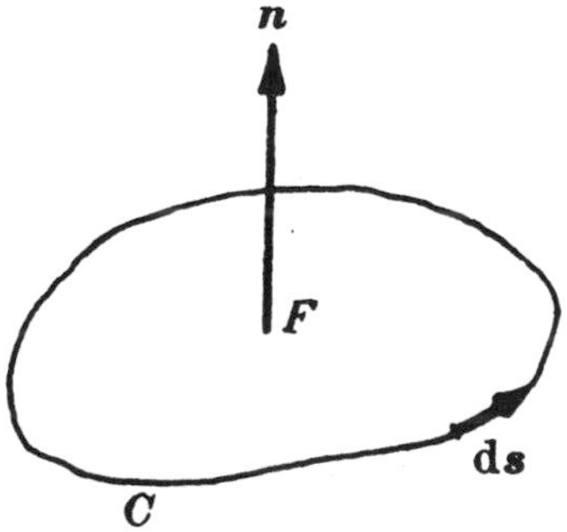

Figure 19.1

the most general form of this law is

$$\oint_C E_s \, \mathrm{d}s = -\frac{1}{c}\frac{\mathrm{d}}{\mathrm{d}t}\int_F H_n \, \mathrm{d}f . \qquad [19.1]$$

In contrast to the static and steady-state cases, the line integral of the electric field intensity over a closed path no longer vanishes. The minus sign in Eq. [19.1] can be interpreted in terms of energy: switching a current on requires the expenditure of work (see Sec. 20).

Equation [19.1] can be transformed into

$$\int_F \mathrm{curl}_n \boldsymbol{E} \, \mathrm{d}f = -\frac{1}{c}\frac{\mathrm{d}}{\mathrm{d}t}\int_F H_n \, \mathrm{d}f = -\frac{1}{c}\int_F \frac{\partial H_n}{\partial t} \, \mathrm{d}f .$$

Since this holds for arbitrary surfaces, it must be true that

$$\mathrm{curl}\, \boldsymbol{E} = -\frac{1}{c}\frac{\partial \boldsymbol{H}}{\partial t} . \qquad [19.2]$$

This is the differential form of Faraday's law of induction.

For *magnetizable bodies*, **B** appears in the place of **H**:

$$\oint_C E_s \, \mathrm{d}s = -\frac{1}{c}\frac{\mathrm{d}}{\mathrm{d}t}\int_F B_n \, \mathrm{d}f , \qquad [19.3]$$

$$\mathrm{curl}\, \boldsymbol{E} = -\frac{1}{c}\frac{\partial \boldsymbol{B}}{\partial t} . \qquad [19.4]$$

From Eq. [19.4] it follows that

$$\frac{\partial}{\partial t} \mathrm{div} \boldsymbol{B} = 0 .$$

This agrees with our stronger postulate of Eq. [12.3],

$$\mathrm{div}\, \boldsymbol{B} = 0 , \qquad [19.5]$$

or, for vacuum,

$$\mathrm{div} \boldsymbol{H} = 0 . \qquad [19.6]$$

If, at any given time, Eq. [19.5] or [19.6] is satisfied, then it follows from Faraday's law that it is satisfied for all times.

Furthermore, Eqs. [3.4] and [16.12] are valid. Thus,

$$\operatorname{div} \boldsymbol{E} = 4\pi\varrho \tag{19.7}$$

and

$$\operatorname{curl} \boldsymbol{B} = \frac{4\pi}{c}\boldsymbol{i}\,, \tag{19.8}$$

or, for vacuum,

$$\operatorname{curl} \boldsymbol{H} = \frac{4\pi}{c}\boldsymbol{i}\,. \tag{19.9}$$

The last two equations are approximations since from them it follows that $\operatorname{div} \boldsymbol{i} = 0$, which condition, however, is not fulfilled exactly. All other equations in this section are exact.

20. THE ENERGY OF CURRENT SYSTEMS

We are now in a position to calculate the energy of a current system. The energy is equal to the work done by the induced current after the applied voltage is switched off. Since the force per unit volume is given by Eqs. [17.5], the work done by the system per unit time and volume is

$$A = \boldsymbol{v}\cdot(\boldsymbol{k}_e + \boldsymbol{k}_m) = \boldsymbol{i}\cdot\boldsymbol{E} + 0\,. \tag{20.1}$$

We assume that the external field is switched off at $t = 0$. Then, the work subsequently done by the induced field is, with the use of Eq. [19.9],

$$W = \int \mathrm{d}t \int A\,\mathrm{d}V = \int_0^\infty \mathrm{d}t \int \boldsymbol{i}\cdot\boldsymbol{E}\,\mathrm{d}V = \frac{c}{4\pi}\int_0^\infty \mathrm{d}t \int \boldsymbol{E}\cdot\operatorname{curl}\boldsymbol{H}\,\mathrm{d}V\,.$$

Because of

$$\boldsymbol{E}\cdot\operatorname{curl}\boldsymbol{H} - \boldsymbol{H}\cdot\operatorname{curl}\boldsymbol{E} = -\operatorname{div}\boldsymbol{E}\times\boldsymbol{H}$$

and Eq. [19.2], we have

$$W = \frac{c}{4\pi}\int_0^\infty \mathrm{d}t \int \boldsymbol{H}\cdot \operatorname{curl}\boldsymbol{E}\,\mathrm{d}V = -\frac{1}{4\pi}\int_0^\infty \mathrm{d}t\int \boldsymbol{H}\cdot\dot{\boldsymbol{H}}\,\mathrm{d}V$$

$$= -\frac{1}{8\pi}\int \boldsymbol{H}^2\,\mathrm{d}V\Big|_{t=0}^{t=\infty}.$$

(The surface integral can, in the quasi-static case, be neglected since the fields at infinity vanish sufficiently fast.) Thus, since $\boldsymbol{H}=0$ at $t=\infty$, we obtain

$$W = \frac{1}{8\pi}\int \boldsymbol{H}^2\,\mathrm{d}V. \qquad [20.2]$$

This is the energy of a current system. This is also equal to the work which must be done in order to switch on the current. The energy W is positive because of the minus sign in Faraday's law of Eq. [19.1].

Since $\operatorname{div}\boldsymbol{H}=0$, it follows that

$$\boldsymbol{H} = \operatorname{curl}\boldsymbol{A} \qquad [20.3]$$

quite rigorously. On the contrary, it follows from Eq. [19.9] that

$$\operatorname{div}\boldsymbol{i} = 0, \qquad [20.4]$$

and, as shown in Section 15, that

$$\boldsymbol{A} = \frac{1}{c}\int\frac{\boldsymbol{i}\,\mathrm{d}V}{r} \qquad [20.5]$$

and

$$\operatorname{div}\boldsymbol{A} = 0. \qquad [20.6]$$

These last three equations are only approximations for slowly varying fields. The energy W given by Eq. [20.2] can be transformed as follows:

$$W = \frac{1}{8\pi}\int \boldsymbol{H}^2\,\mathrm{d}V = \frac{1}{8\pi}\int \boldsymbol{H}\cdot\operatorname{curl}\boldsymbol{A}\,\mathrm{d}V.$$

After partial integration, dropping the surface integral, and using Eq. [19.9],

$$W = \frac{1}{8\pi}\int \boldsymbol{A}\cdot \operatorname{curl}\boldsymbol{H}\,\mathrm{d}V = \frac{1}{2c}\int \boldsymbol{A}\cdot\boldsymbol{i}\,\mathrm{d}V\,.$$

Thus, with Eq. [20.5],

$$W = \frac{1}{2c^2}\iint \frac{\boldsymbol{i}_P\cdot\boldsymbol{i}_{P'}}{r_{PP'}}\,\mathrm{d}V_P\,\mathrm{d}V_{P'}\,. \qquad [20.7]$$

Equation [20.7] exhibits a definite analogy to the energy of the electric field given in Eq. [4.5]. The expression of Eq. [20.7] can be written for individual circuits as follows.

a. One circuit. Here,

$$W = \tfrac{1}{2}\,LJ^2\,. \qquad [20\ 8]$$

The quantity L, called the *coefficient of self-inductance*, is defined so that

$$LJ^2 = \frac{1}{c^2}\iint \frac{\boldsymbol{i}_P\cdot\boldsymbol{i}_{P'}}{r_{PP'}}\,\mathrm{d}V_P\,\mathrm{d}V_{P'}\,. \qquad [20.9]$$

This is reasonable, since L depends only upon the geometry of the circuit. That is, if the current J is multiplied by some factor, then $\boldsymbol{i}$ at every point is also multiplied by the same factor. Here, it is only presupposed that the dc current distribution can be used in the integrand. For rapidly varying fields, L becomes frequency dependent.

b. Two circuits. In this case the energy of the magnetic field takes the positive definite quadratic form

$$W = \tfrac{1}{2}\,L_1 J_1^2 + L_{12} J_1 J_2 + \tfrac{1}{2}\,L_2 J_2^2\,. \qquad [20.10]$$

Here,

$$L_{12} J_1 J_2 = \frac{1}{c^2}\int_1 \mathrm{d}V_P \int_2 \mathrm{d}V_{P'}\,\frac{\boldsymbol{i}_P\cdot\boldsymbol{i}_{P'}}{r_{PP'}}\,. \qquad [20.11]$$

If the conductors are filamentary, then

$$L_{12} J_1 J_2 = J_1 J_2\,\frac{1}{c^2}\oint_1\oint_2 \frac{\mathrm{d}\boldsymbol{s}_1\cdot\mathrm{d}\boldsymbol{s}_2}{r_{12}}\,. \qquad [20.12]$$

This transition is allowed since $r_{PP'}$ remains finite. The *coefficient of mutual inductance* then becomes

$$L_{12} = L_{21} = \frac{1}{c^2}\oint_1\oint_2 \frac{\mathbf{ds}_1 \cdot \mathbf{ds}_2}{r_{12}}. \qquad [20.13]$$

On the other hand, the transition from spatial to filamentary conductors cannot be made for the case of self-inductance; otherwise, completely erroneous conclusions may result.

The generalization to more than two circuits is obvious.

Coefficient of self-inductance for a solenoid. According to Eq. [15.5],

$$H = \frac{4\pi N}{h}\frac{J}{c}.$$

Thus,

$$\frac{1}{2}LJ^2 = \frac{1}{8\pi}\int H^2\,\mathrm{d}V = \frac{1}{8\pi}\pi a^2 h \frac{16\pi^2 N^2}{h^2}\frac{J^2}{c^2},$$

$$L = \frac{1}{c^2}\frac{4\pi^2 a^2 N^2}{h} = \frac{1}{c^2}\frac{(2\pi a N)^2}{h} = \frac{1}{c^2}\frac{\Lambda^2}{h}, \qquad [20.14]$$

where $\Lambda = 2\pi a N$ is the total length of the winding (see Fig. 20.1).

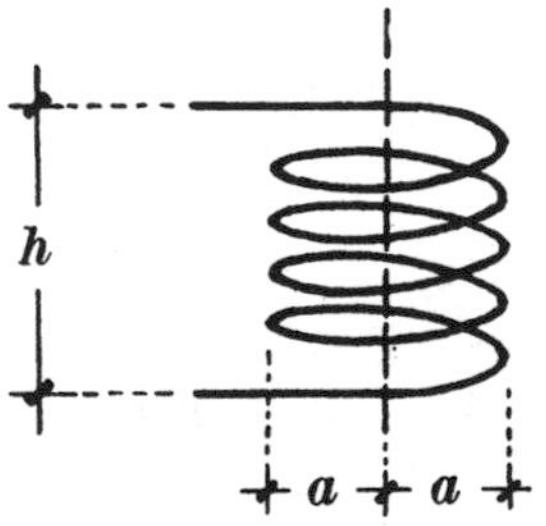

Figure 20.1

Dimensions of self-inductance. In cgs units,

$$L_{\text{emu}} = c^2 L_{\text{esu}}, \qquad [L_{\text{emu}}] = \text{cm}.$$

In the volt-ampere (rationalized mks) system,

$$1 \text{ henry} = 1 \frac{\text{volt} \cdot \text{sec}}{\text{amp}} = 10^9 \text{ emu} .$$

21. TIME-DEPENDENT CURRENT FLOW IN CIRCUITS

From Eqs. [19.2] and [20.3] it follows that

$$\operatorname{curl}\left(\boldsymbol{E} + \frac{1}{c}\frac{\partial \boldsymbol{A}}{\partial t}\right) = 0 .$$

We thus have the following generalization of the electrostatic relation of Eq. [2.3]:

$$\boldsymbol{E} = -\frac{1}{c}\frac{\partial \boldsymbol{A}}{\partial t} - \operatorname{grad} \varphi . \qquad [21.1]$$

We now wish to formulate an important relation for the variation of the current in a circuit. Let the current distribution in space be given. We assume that it is the same

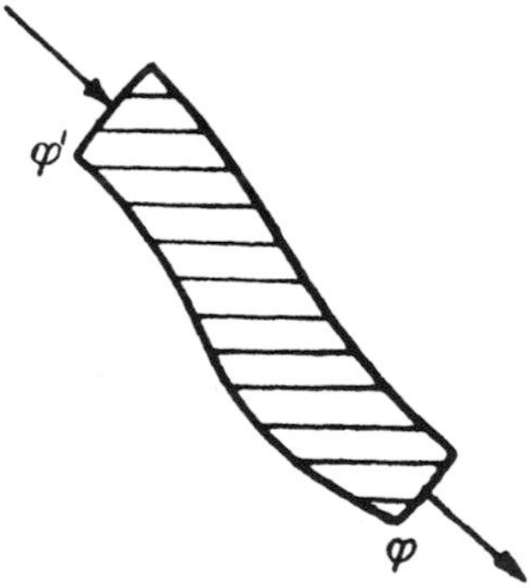

Figure 21.1

as for a steady-state current. We first consider an ***ordinary conductor with two electrodes***. As demonstrated in Section 14,

$$\boldsymbol{i} = \sigma \boldsymbol{E} , \qquad [21.2]$$

while the heat produced per unit time is, on the one hand,

given by

$$\bar{Q} = \int \frac{\boldsymbol{i}^2}{\sigma} \, \mathrm{d}V = J^2 R \qquad [21.3]$$

from the definition of the resistance R, and, on the other hand, by

$$\bar{Q} = \int \boldsymbol{i} \cdot \boldsymbol{E} \, \mathrm{d}V = -\int \boldsymbol{i} \cdot \operatorname{grad} \varphi \, \mathrm{d}V - \frac{1}{c} \int \boldsymbol{i} \cdot \dot{\boldsymbol{A}} \, \mathrm{d}V$$

because of Eq. [21.1]. If we neglect $\operatorname{div} \boldsymbol{i}$ in

$$\operatorname{div}(\varphi \boldsymbol{i}) = \varphi \operatorname{div} \boldsymbol{i} + \boldsymbol{i} \cdot \operatorname{grad} \varphi$$

and substitute $\boldsymbol{A}$ from Eq. [20.5], we get

$$\bar{Q} = -\oint \varphi i_n \, \mathrm{d}f - \frac{1}{c^2} \iint \frac{\boldsymbol{i}_P \cdot \dfrac{\partial \boldsymbol{i}_{P'}}{\partial t}}{r_{PP'}} \, \mathrm{d}V_P \, \mathrm{d}V_{P'} \, .$$

Differentiating Eq. [20.9] with respect to time results in

$$2LJ\dot{J} = 2 \frac{1}{c^2} \iint \frac{\boldsymbol{i}_P \cdot \dfrac{\partial \boldsymbol{i}_{P'}}{\partial t}}{r_{PP'}} \, \mathrm{d}V_P \, \mathrm{d}V_{P'}$$

(since interchanging P and P' does not affect the integral). Thus, we obtain

$$\bar{Q} = J^2 R = -\oint \varphi i_n \, \mathrm{d}f - LJ\dot{J} \, , \qquad J^2 R = \varepsilon J - LJ\dot{J} \, ,$$

where

$$\varepsilon = \varphi' - \varphi$$

is the *electromotive force*. We thus obtain the important equation

$$L\dot{J} + RJ = \varepsilon \, . \qquad [21.4]$$

This could have been obtained more easily by assuming a filamentary conductor. The latter procedure, however, is not correct.

For a *circuit containing a capacitor*, an additional term is obtained:

$$-\oint \varphi i_n \,\mathrm{d}f = \varepsilon J - (\varphi_{1'} - \varphi_{2'})J\,.$$

The sum of the charges on 1′ and 2′ (Fig. 21.2) remains constant and equal to zero. What flows in at the left

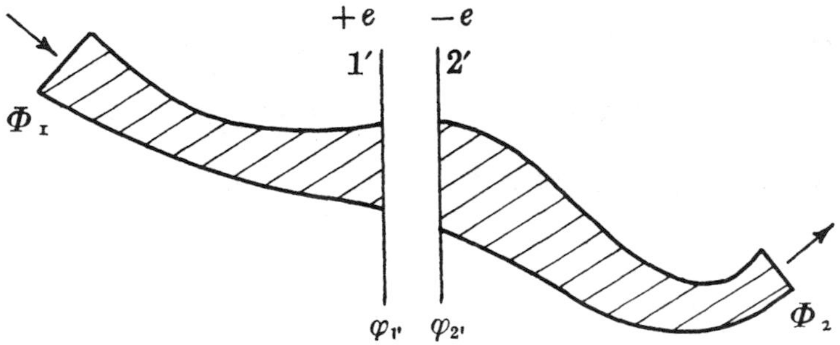

Figure 21.2

flows out at the right; it is as if the current flowed without interruption. If the charge is denoted by e and the capacitance by C, then

$$\varphi_{1'} - \varphi_{2'} = \frac{e}{C}\,.$$

On the other hand,

$$\dot{e} = J\,.$$

In this case the differential equation becomes

$$L\dot{J} + WJ + \frac{e}{C} = \varepsilon \qquad [21.5]$$

or, differentiated,

$$L\ddot{J} + W\dot{J} + \frac{J}{C} = \dot{\varepsilon}\,. \qquad [21.6]$$

For the case of *two circuits*, Q is calculated separately for each. However, $\boldsymbol{A}$ is now due to both circuits. Thus,

$$\bar{Q}_1 = \int_1 \boldsymbol{i}\cdot\boldsymbol{E}\,\mathrm{d}V = J_1^2 R_1 = -\oint_1 \varphi i_n\,\mathrm{d}f - \frac{1}{c}\int_1 \boldsymbol{i}\cdot\dot{\boldsymbol{A}}\,\mathrm{d}V\,,$$

$$\boldsymbol{A}_P = \frac{1}{c}\int_1 \frac{\boldsymbol{i}_{P'}}{r_{PP'}}\,\mathrm{d}V_{P'} + \frac{1}{c}\int_2 \frac{\boldsymbol{i}_{P'}}{r_{PP'}}\,\mathrm{d}V_{P'}\,.$$

Analogous to the preceding calculation one obtains

$$\begin{aligned} L_1\dot{J}_1 + L_{12}\dot{J}_2 + R_1 J_1 + \frac{e_1}{C_1} &= \varepsilon_1\,,\\ L_2\dot{J}_2 + L_{12}\dot{J}_1 + R_2 J_2 + \frac{e_2}{C_2} &= \varepsilon_2\,. \end{aligned} \qquad [21.7]$$

The terms e_i/C_i drop out if there are no capacitors in the circuit. The energy is

$$E = \frac{1}{2}L_1 J_1^2 + L_{12}J_1 J_2 + \frac{1}{2}L_2 J_2^2 + \frac{1}{2}\frac{e_1^2}{C_1} + \frac{1}{2}\frac{e_2^2}{C_2}\,. \qquad [21.8]$$

Because of

$$J_1 = \dot{e}_1\,, \qquad J_2 = \dot{e}_2\,,$$

it follows that

$$\frac{\mathrm{d}E}{\mathrm{d}t} = \varepsilon_1 J_1 + \varepsilon_2 J_2 - R_1 J_1^2 - R_2 J_2^2\,, \qquad [21.9]$$

which is equal to the work done by external forces minus the heat produced per unit time. In the above differential equations, R has the character of a frictional force since the equations are not invariant when J is replaced by $-J$ and t by $-t$.

We have obtained ordinary differential equations rather than partial differential equations because we have assumed that the current distribution is known, and also because radiative effects have been neglected.

Integration of the differential equations for some interesting cases

1. *One circuit without capacitance* (*an RL circuit*). The differential equation is

$$L\dot{J} + RJ = \varepsilon\,. \qquad [21.10]$$

Switching off the emf,

$$\begin{cases} \varepsilon = \varepsilon_0 & \text{for} \quad t < 0\,, \\ \varepsilon = 0 & \text{for} \quad t > 0\,, \end{cases}$$

gives the solution

$$\text{for} \quad t < 0: \quad J = J_0 = \frac{\varepsilon_0}{R} \quad \text{(Ohm's law)}\,,$$

$$\text{for} \quad t > 0: \quad \dot{J} = -\frac{R}{L} J\,,$$

$$J = J_0 e^{-\frac{R}{L}t}\,.$$

(The constant of integration has been chosen so that the current is continuous at $t=0$.) This solution corresponds to a gradual decay of the current.

Switching on the emf,

$$\begin{cases} \varepsilon = 0 & \text{for} \quad t < 0\,, \\ \varepsilon = \varepsilon_0 & \text{for} \quad t > 0\,, \end{cases}$$

implies

$$\text{for} \quad t < 0: \quad J = 0\,,$$

$$\text{for} \quad t > 0: \quad \dot{J} + \frac{R}{L}\left(J - \frac{\varepsilon_0}{R}\right) = 0\,,$$

$$J - \frac{\varepsilon_0}{R} = C \cdot e^{-\frac{R}{L}t}\,,$$

$$C = -\frac{\varepsilon_0}{R}\,,$$

so that J is continuous at $t=0$:

$$J = \frac{\varepsilon_0}{R}\left(1 - e^{-\frac{R}{L}t}\right).$$

Thus $J \rightarrow \varepsilon_0/R$ as $t \rightarrow \infty$. Initially, the current increases gradually because of the self-inductance. The final current is established in a time of the order of magnitude of L/R (proportional to $1/R$).

2. *Applied electromotive force with periodic time dependence.* For linear expressions it is convenient to use complex exponential notation. The real (or imaginary) part can be taken later, as required. That is, if we find a complex solution of a linear differential equation with real coefficients, then both the real and the imaginary parts are also solutions.

We therefore assume that

$$\varepsilon = \varepsilon_0 e^{i\omega t},$$

where ω is the angular frequency, equal to 2π times the frequency and try

$$J = A e^{i\omega t}.$$

For this type of solution the differential operations are equivalent to certain multiplications:

$$\frac{\mathrm{d}}{\mathrm{d}t} \sim i\omega\,,$$

$$\frac{\mathrm{d}^2}{\mathrm{d}t^2} \sim (i\omega)^2 = -\,\omega^2\,.$$

Equation [21.10] then yields

$$(Li\omega + R)A = \varepsilon_0\,,$$

$$A = \frac{\varepsilon_0}{R + i\omega L} = \frac{\varepsilon_0(R - i\omega L)}{R^2 + \omega^2 L^2}\,.$$

If $\varepsilon_0 = |\varepsilon_0| e^{i\alpha}$, then we can also write

$$A = |A|\, e^{i(\alpha+\vartheta)}\,,$$

where

$$|A| = |\varepsilon_0|/(R^2+\omega^2L^2)^{\frac{1}{2}} \text{ and } \tan\vartheta = -\omega L/R.$$

The phase of the current is shifted by an amount ϑ with respect to the phase of the applied electromotive force, since

$$\mathrm{Re}\,(\varepsilon) = |\varepsilon_0|\cos(\omega t + \alpha)$$

and

$$\mathrm{Re}\,(J) = |A|\cos(\omega t + \alpha + \vartheta)\,.$$

3. *One circuit with capacitance (an RLC circuit).* The differential equation is

$$L\ddot{J} + R\dot{J} + \frac{1}{C}J = \dot{\varepsilon}\,. \qquad [21.11]$$

a. Forced oscillations: $\varepsilon = \varepsilon_0 e^{i\omega t}$. We assume that $\alpha = 0$ and that ε_0 is real and positive and try

$$J = Ae^{i\omega t}\,.$$

Thus, Eq. [21.11] yields

$$\left(-\omega^2 L + i\omega R + \frac{1}{C}\right)A = \varepsilon_0 i\omega\,,$$

$$A = \frac{i\omega\varepsilon_0}{-\omega^2 L + i\omega R + 1/C} = |A|\,e^{i\vartheta}\,,$$

where

$$|A| = \frac{\varepsilon_0}{\sqrt{R^2 + (\omega L - 1/C\omega)^2}}\,, \qquad \tan\vartheta = -\frac{1}{R}\left(\omega L - \frac{1}{C\omega}\right).$$

The maximum of $|A|$ occurs when the denominator is minimum:

$$\omega_m L - \frac{1}{C\omega_m} = 0\,,$$

$$\left.\begin{aligned} \omega_m &= \frac{1}{\sqrt{LC}} \\ |A|_{\max} &= \frac{\varepsilon_0}{R} \end{aligned}\right\} \quad \textit{resonance.}$$

The quantity $\tan\vartheta$ changes sign when going through the resonance.

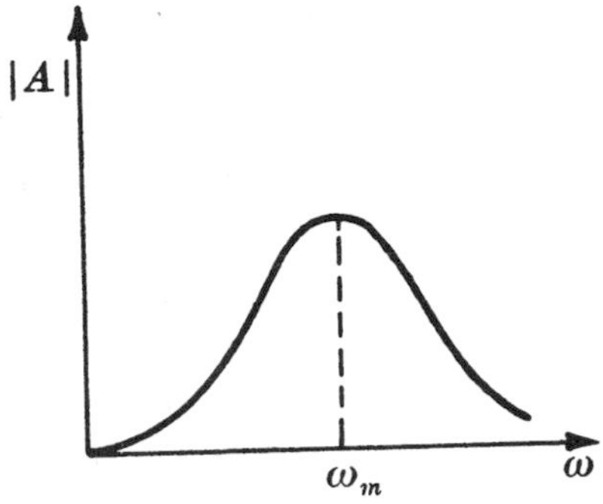

Figure 21.3

b. Free oscillations: $\varepsilon = 0$. Put

$$J = Ae^{i\omega_0 t}.$$

The oscillation frequency ω_0 is now not established by the emf but, instead, is determined by the differential equation of [21.11]:

$$-L\omega_0^2 + Ri\omega_0 + \frac{1}{C} = 0\,, \qquad \omega_0 = \frac{iR}{2L} \pm \sqrt{\frac{1}{LC} - \frac{R^2}{4L^2}}\,.$$

There are two possibilities: [1]

(α) $$\frac{R^2}{4L^2} < \frac{1}{LC}\,, \qquad R < 2\sqrt{\frac{L}{C}}\,;$$

$$\mathrm{Re}\,(J) = Ae^{-\frac{R}{2L}t} \cdot \cos\left(\sqrt{\frac{1}{LC} - \frac{R^2}{4L^2}}\right)t$$

[1] *Translator's Note*: In addition to the damped and overdamped cases discussed in α and β, there also exists a third possibility, namely, when $R^2/4L^2 = 1/LC$. In this case,

$$\omega_0 = iR/2L = i/\sqrt{LC}$$

and

$$\mathrm{Re}(J) = Ae^{-(R/2L)t} = Ae^{-t/\sqrt{LC}} = Ae^{-\omega_m t}\,.$$

This is called the *critically damped* case.

(with appropriate choice of the time origin). This corresponds to *damped oscillations.*

$$(\beta) \qquad \frac{R^2}{4L^2} > \frac{1}{LC}, \qquad R > 2\sqrt{\frac{L}{C}};$$

$$\omega_0 = i\left(\frac{R}{2L} \pm \sqrt{\frac{R^2}{4L^2} - \frac{1}{LC}}\right),$$

$$\operatorname{Re}(J) = A \cdot e^{-\left(\frac{R}{2L} \pm \sqrt{\frac{R^2}{4L^2} - \frac{1}{LC}}\right)t}.$$

Here, there are two solutions, corresponding to both signs of the square root. This is the *aperiodic* (*overdamped*) case.

In the periodic case α, the oscillation frequency ω_0 is somewhat smaller than the resonance frequency ω_m.

If $R \ll 2\sqrt{L/C}$, then

$$\omega_0 \to \omega_m = 1/\sqrt{LC},$$

which is sometimes called the Thompson equation.

4. *Two RLC circuits in parallel.* If the mutual interaction between circuits 1 and 2 is neglected, Eq. [21.7] differ-

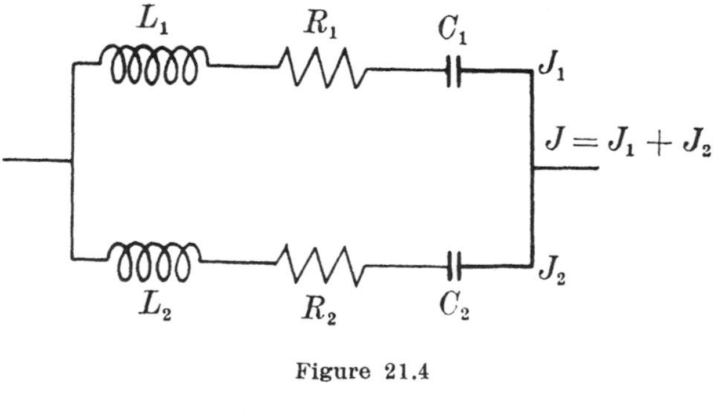

Figure 21.4

entiated with respect to time yields

$$\dot{\varepsilon} = L_1\ddot{J}_1 + R_1\dot{J}_1 + \frac{J_1}{C_1} = L_2\ddot{J}_2 + R_2\dot{J}_2 + \frac{J_2}{C_2}.$$

In the spirit of forced oscillations we set

$$J_1 = A_1 e^{i\omega t},$$
$$J_2 = A_2 e^{i\omega t},$$
$$\varepsilon = \varepsilon_0 e^{i\omega t}.$$

Then,

$$i\omega\varepsilon_0 = A_1\left(-L_1\omega^2 + R_1 i\omega + \frac{1}{C_1}\right) = A_2\left(-L_2\omega^2 + R_2 i\omega + \frac{1}{C_2}\right),$$

$$A_1 + A_2 = i\omega\varepsilon_0\left\{\frac{1}{-L_1\omega^2 + R_1 i\omega + 1/C_1} + \frac{1}{-L_2\omega^2 + R_2 i\omega + 1/C_2}\right\},$$

$$J = i\omega\varepsilon\left\{\frac{1}{-L_1\omega^2 + R_1 i\omega + 1/C_1} + \frac{1}{-L_2\omega^2 + R_2 i\omega + 1/C_2}\right\}.$$

For the case of direct current, in the absence of capacitors,

$$J = \varepsilon\left(\frac{1}{R_1} + \frac{1}{R_2}\right).$$

22. THE SKIN EFFECT

Until now, we have assumed that the current distributions are the same as for the direct current case. However, with increasing frequency the self-inductance tends to restrict the current to the surfaces of the conductor. The resistance is thereby increased since the effective cross-sectional area is diminished.

We will perform the calculation for a semi-infinite conductor. (Bessel functions appear for cylindrical conductors.)

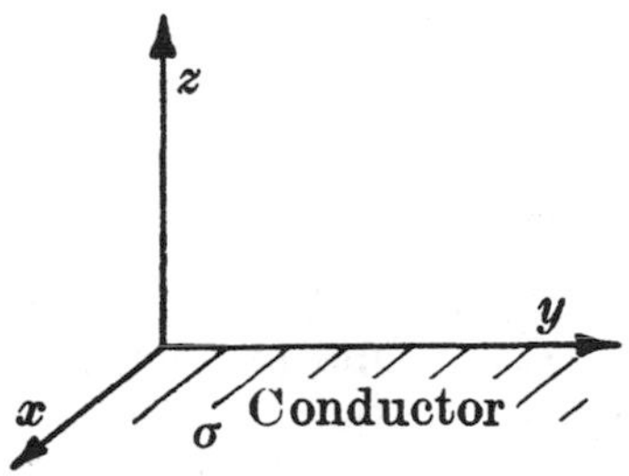

Figure 22.1

Let the current distribution be parallel to the y axis. (For direct current, it would be uniform.) We employ the equations

$$\operatorname{curl} \boldsymbol{H} = \frac{4\pi}{c} \boldsymbol{i}, \qquad [22.1]$$

$$\operatorname{curl} \boldsymbol{E} = -\frac{1}{c} \dot{\boldsymbol{H}}, \qquad [22.2]$$

$$\boldsymbol{i} = \sigma \boldsymbol{E}, \qquad [22.3]$$

$$\operatorname{div} \boldsymbol{H} = 0. \qquad [22.4]$$

Because of Eqs. [22.3] and [20.4], it is also true that

$$\operatorname{div} \boldsymbol{E} = 0. \qquad [22.5]$$

From the first three equations it follows that

$$\operatorname{curl} \operatorname{curl} \boldsymbol{E} = -\frac{4\pi}{c^2} \frac{\partial \boldsymbol{i}}{\partial t} = -\frac{4\pi}{c^2} \sigma \frac{\partial \boldsymbol{E}}{\partial t},$$

$$\operatorname{curl} \operatorname{curl} \boldsymbol{H} = -\frac{4\pi}{c^2} \sigma \frac{\partial \boldsymbol{H}}{\partial t}.$$

Because of the identity given in Eq. [15.9], consideration of [22.5] and [19.6] leads to

$$\nabla^2 \boldsymbol{E} = \frac{4\pi}{c^2} \sigma \frac{\partial \boldsymbol{E}}{\partial t},$$
$$\nabla^2 \boldsymbol{H} = \frac{4\pi}{c^2} \sigma \frac{\partial \boldsymbol{H}}{\partial t}. \qquad [22.6]$$

These two relations are very similar to the equation for heat conduction. They are not exact because the usual approximations for quasi-static fields have been made. Nevertheless, they suffice to explain the skin effect. We will assume that the fields have the form $E_y(z)$, $H_x(z)$, all other components being zero. This hypothesis is sufficient. Let the fields be periodic in time:

$$E_y = E_y^0(z)\, e^{i\omega t}.$$

Substituting into [22.6],

$$\frac{d^2 E_y^0}{dz^2} = \frac{4\pi\sigma\omega}{c^2} i E_y^0 .$$

With $k^2 = 4\pi\sigma\omega i/c^2$, the solution is

$$E_y^0 = A e^{kz} .$$

Introducing the quantity $\varkappa = \sqrt{2\pi\sigma\omega}/c$, so that

$$k^2 = \varkappa^2 2i , \qquad k = \pm \varkappa(1 + i) , \tag{22.7}$$

the solution becomes

$$E_y = A e^{\varkappa z} e^{i(\varkappa z + \omega t)}$$

and

$$\operatorname{Re} E_y = A e^{\varkappa z} \cos(\omega t + \varkappa z) . \tag{22.8}$$

We must restrict ourselves to the positive value of k since E_y cannot increase with increasing depth. Thus, E_y *decreases exponentially as* $z \to -\infty$. (For $\omega = k = 0$, one obtains at most a linear dependence of E_y on z.)

The effective depth to which the alternating current penetrates, $1/\varkappa$, is

$$\frac{1}{\varkappa} = \frac{1}{\sqrt{2\pi\sigma_{\mathrm{emu}}\omega}} , \tag{22.9}$$

since

$$\sigma/c^2 = \sigma_{\mathrm{emu}} .$$

Numerical estimate. For a frequency $\nu = \omega/2\pi = 50\ \mathrm{sec}^{-1}$, for copper where $\sigma_{\mathrm{emu}} = 5.9 \cdot 10^{-4}\ \mathrm{sec \cdot cm^{-2}}$,

$$1/\varkappa \sim 1\ \mathrm{cm} .$$

For the resistance of a cylindrical wire, a calculation yields the asymptotic formula ($\varkappa a \gg 1$)

$$R = R_0 \frac{\varkappa a}{2} .$$

Here, R_0 is the dc resistance and a the radius of the wire.

23. THE LAW OF INDUCTION FOR MOVING CONDUCTORS

The following considerations represent a preliminary step toward the theory of relativity.

Let the conductor move with velocity $\boldsymbol{v}$. The force on a moving charge is not $e\boldsymbol{E}$ but, according to Eq. [17.6],

$$e\left(\boldsymbol{E}+\frac{1}{c}\,\boldsymbol{v}\times\boldsymbol{H}\right).$$

For a moving conductor, Ohm's law must be modified to the extent that

$$\boldsymbol{E}^*=\boldsymbol{E}+\frac{1}{c}\,(\boldsymbol{v}\times\boldsymbol{H}) \qquad [23.1]$$

appears in place of $\boldsymbol{E}$, so that the current density is given by

$$\boldsymbol{i}=\sigma\boldsymbol{E}^*\,. \qquad [23.2]$$

(The above is correct only to first order in v/c. An exact consideration shows that σ is dependent upon direction.)

Now, purely mathematically, without any additional physical assumptions, it can be shown that if

$$\oint E_s\,\mathrm{d}s=-\frac{1}{c}\frac{\mathrm{d}}{\mathrm{d}t}\int H_n\,\mathrm{d}f \qquad [23.3]$$

is valid for a stationary surface, then

$$\oint_{\rightarrow} E_s^*\,\mathrm{d}s=-\frac{1}{c}\frac{\mathrm{d}}{\mathrm{d}t}\int_{\rightarrow} H_n\,\mathrm{d}f \qquad [23.4]$$

holds for a moving surface. (The symbol $\rightarrow$ is to denote that the surface and its bounding curve are in motion.)

This, together with the physical law of Eq. [23.2], indicates that *for the same relative motion between conductor and magnet, the same current is induced, whether* (*a*) *the magnet is moved* *slowly with respect to the stationary conductor* (Fig. 23.1*a*), *or* (*b*) *the conductor is moved with the same—but oppositely directed—velocity with respect to the stationary magnet* (Fig. 23.1*b*), since $\oint \boldsymbol{E}_s^* \mathrm{d}\boldsymbol{s}$ is the same in both cases. This

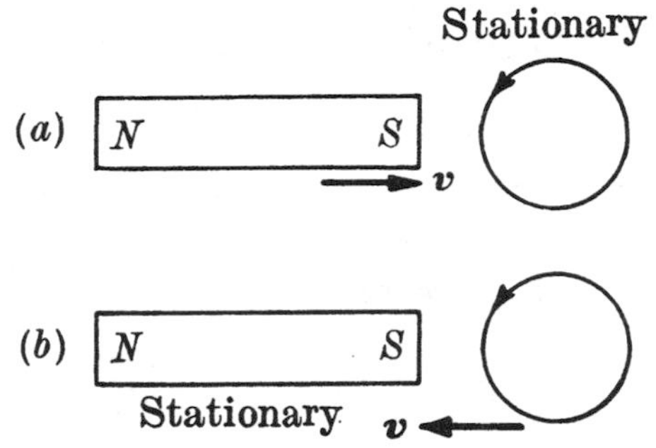

Figure 23.1

fact is more extensively generalized in the theory of relativity.

Proof of this mathematical theorem. For an arbitrary vector $\boldsymbol{B}$ at the time t_0, it is true that

$$\frac{\mathrm{d}}{\mathrm{d}t}\int_{\rightarrow} B_n \,\mathrm{d}f = \int \frac{\partial B_n}{\partial t}\,\mathrm{d}f + \frac{\mathrm{d}}{\mathrm{d}t}\int_{t} B_n(t_0)\,\mathrm{d}f\,. \qquad [23.5]$$

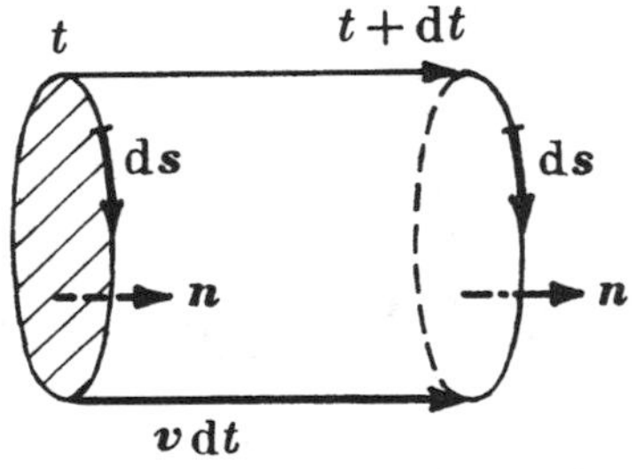

Figure 23.2

We now construct a closed surface by supplementing the surface at times t and $t+\mathrm{d}t$ with the area of the cylinder swept out by the conductor during the time $\mathrm{d}t$. On this cylindrical surface,

$$\mathrm{d}\boldsymbol{f} = \mathrm{d}\boldsymbol{s} \times \boldsymbol{v}\, \mathrm{d}t\,.$$

Applying Gauss's theorem to the closed surface gives

$$\int \operatorname{div} \boldsymbol{B}\, \mathrm{d}V = \int_{t+\mathrm{d}t} B_n(t)\, \mathrm{d}f - \int_{t} B_n(t)\, \mathrm{d}f + \int_{C} \boldsymbol{B} \cdot (\mathrm{d}\boldsymbol{s} \times \boldsymbol{v}\, \mathrm{d}t)\,,$$

where the argument of $\boldsymbol{B}$ is evaluated at a fixed time t. Transforming the last integral by means of Stokes's theorem and using $\mathrm{d}V = \mathrm{d}f v_n \mathrm{d}t$ results in

$$\int_{t+\mathrm{d}t} B_n(t)\, \mathrm{d}f - \int_{t} B_n(t)\, \mathrm{d}f = \int \operatorname{div} \boldsymbol{B}\, \mathrm{d}V - \mathrm{d}t \int_{C} \mathrm{d}\boldsymbol{s} \cdot (\boldsymbol{v} \times \boldsymbol{B})$$

$$= \mathrm{d}t \int \{\boldsymbol{v} \operatorname{div} \boldsymbol{B} - \operatorname{curl}(\boldsymbol{v} \times \boldsymbol{B})\}_n\, \mathrm{d}f\,.$$

Substituting this into Eq. [23.5] yields the *mathematical identity*

$$\frac{\mathrm{d}}{\mathrm{d}t} \int_{\rightarrow} B_n\, \mathrm{d}f = \int \left\{ \frac{\partial \boldsymbol{B}}{\partial t} + \boldsymbol{v} \operatorname{div} \boldsymbol{B} - \operatorname{curl}(\boldsymbol{v} \times \boldsymbol{B}) \right\}_n \mathrm{d}f\,. \qquad [23.6]$$

This relation is valid for any arbitrary differentiable vector $\boldsymbol{B}$.

We now apply the above to the case $\boldsymbol{B} = \boldsymbol{H}$. It is essential to use the fact that

$$\operatorname{div} \boldsymbol{H} = 0\,.$$

There then remains

$$-\frac{1}{c} \frac{\mathrm{d}}{\mathrm{d}t} \int_{\rightarrow} H_n\, \mathrm{d}f = -\frac{1}{c} \int \left(\frac{\partial \boldsymbol{H}}{\partial t} \right)_n \mathrm{d}f + \frac{1}{c} \int (\boldsymbol{v} \times \boldsymbol{H})_s\, \mathrm{d}s\,,$$

which, because of Eq. [19.2] and Stokes's theorem, becomes

$$\oint \left\{ \boldsymbol{E} + \frac{1}{c} \boldsymbol{v} \times \boldsymbol{H} \right\}_s \mathrm{d}s = \oint E_s^*\, \mathrm{d}s\,.$$

The theorem is thus proved.

Chapter 4. Rapidly Varying Fields

24. THE MAXWELL EQUATIONS

The equations which we have applied to quasi-static fields contain an inconsistency. The equations for vacuum were

$$\operatorname{curl} \boldsymbol{H} = \frac{4\pi}{c} \boldsymbol{i} , \qquad \operatorname{div} \boldsymbol{H} = 0 ,$$

$$\operatorname{curl} \boldsymbol{E} = - \frac{1}{c} \frac{\partial \boldsymbol{H}}{\partial t} , \qquad \operatorname{div} \boldsymbol{E} = 4\pi\varrho .$$

From these it follows that

$$\operatorname{div} \boldsymbol{i} = 0 .$$

This, however, is not exactly true since, in general, we also have the continuity equation

$$\frac{\partial \varrho}{\partial t} + \operatorname{div} \boldsymbol{i} = 0 .$$

(Surface charge densities can always be considered as limiting cases of volume charge densities.) The situation can easily be corrected by adding an additional term. From the fourth equation and the continuity equation it follows that

$$\operatorname{div} \left\{ \boldsymbol{i} + \frac{1}{4\pi} \frac{\partial \boldsymbol{E}}{\partial t} \right\} = 0 .$$

Since $\operatorname{div}\operatorname{curl}\boldsymbol{H} \equiv 0$, our equations become consistent if $\boldsymbol{i}$ is replaced by $\boldsymbol{i}+(1/4\pi)(\partial\boldsymbol{E}/\partial t)$. Thus,

$$\operatorname{curl}\boldsymbol{H} = \frac{4\pi}{c}\boldsymbol{i} + \frac{1}{c}\frac{\partial\boldsymbol{E}}{\partial t}.$$

The additional term $(1/c)(\partial\boldsymbol{E}/\partial t)$, which is called the ***displacement current***, is due to Maxwell. This term vanishes for steady-state fields. We now have the following completely self-consistent set of equations:

$$\operatorname{curl}\boldsymbol{E} = -\frac{1}{c}\dot{\boldsymbol{H}}, \tag{24.1}$$

$$\operatorname{curl}\boldsymbol{H} = \frac{4\pi}{c}\boldsymbol{i} + \frac{1}{c}\dot{\boldsymbol{E}}, \tag{24.2}$$

$$\operatorname{div}\boldsymbol{H} = 0, \tag{24.3}$$

$$\operatorname{div}\boldsymbol{E} = 4\pi\varrho. \tag{24.4}$$

These are the famous ***Maxwell equations for vacuum.***

The continuity equation now appears as a consequence of the Maxwell equations, since from Eqs. [24.2] and [24.4] it follows that

$$\dot{\varrho} + \operatorname{div}\boldsymbol{i} = 0. \tag{24.5}$$

A somewhat analogous result holds for surface charge den-

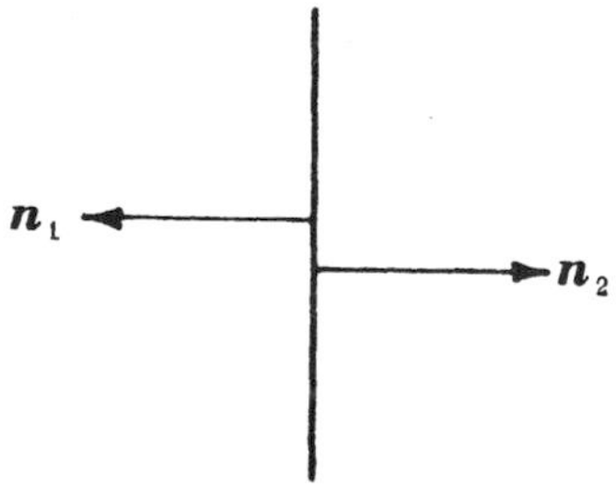

Figure 24.1

sities. According to Eq. [7.3],

$$4\pi\omega = E_{n_1} + E_{n_2}.$$

The conservation law for electrical charge requires that

$$\dot{\omega} + i_{n_2} + i_{n_2} = 0\,.$$

From these two equations it follows that

$$4\pi(i_{n_1} + i_{n_2}) + \dot{E}_{n_1} + \dot{E}_{n_2} = 0\,.$$

That is, the normal component of $4\pi \boldsymbol{i} + \dot{\boldsymbol{E}}$ is continuous across a boundary.

Molecular viewpoint. The above equations are correct for the field intensities in vacuum. They are, therefore, also valid for the microscopic fields $\boldsymbol{e}$ and $\boldsymbol{h}$. Forming the spatial averages of $\boldsymbol{e}$ and $\boldsymbol{h}$ leads to the macroscopic fields $\boldsymbol{E}$ and $\boldsymbol{B}$,

$$\bar{\boldsymbol{e}} = \boldsymbol{E}\,, \qquad \bar{\boldsymbol{h}} = \boldsymbol{B}\,,$$

for which the following equations hold:

$$\operatorname{curl} \boldsymbol{E} = -\frac{1}{c}\dot{\boldsymbol{B}}\,, \tag{24.1'}$$

$$\operatorname{curl} \boldsymbol{B} = \frac{4\pi}{c}\boldsymbol{i} + \frac{1}{c}\dot{\boldsymbol{E}}\,, \tag{24.2'}$$

$$\operatorname{div} \boldsymbol{B} = 0\,, \tag{24.3'}$$

$$\operatorname{div} \boldsymbol{E} = 4\pi\varrho\,. \tag{24.4'}$$

These are the ***Maxwell equations for material media.*** In these equations, ϱ denotes the total charge density (including polarization charges) and $\boldsymbol{i}$ is the total current density,

$$\boldsymbol{i} = \boldsymbol{i}_c + c \operatorname{curl} \boldsymbol{M} + \dot{\boldsymbol{P}}\,, \tag{24.6}$$

where $\boldsymbol{i}_c$ is the conduction current density, $c \operatorname{curl} \boldsymbol{M}$ is the molecular current density, and $\dot{\boldsymbol{P}}$ is the polarization current density. Since the polarization $\boldsymbol{P}$ is equal to the electric dipole moment per unit volume,

$$\boldsymbol{P} = \left(\sum e\boldsymbol{x}\right)_{\text{per unit volume}}\,,$$

then

$$\dot{\boldsymbol{P}} = \left(\sum e\boldsymbol{v}\right)_{\text{per unit volume}}$$

is indeed a current density resulting from the variation of the dipole moment.

The Maxwell equations [24.2′] and [24.4′] can be written in terms of $\boldsymbol{D}$ and $\boldsymbol{H}$, defined, respectively, by

$$\boldsymbol{D} = \boldsymbol{E} + 4\pi\boldsymbol{P} \tag{24.7}$$

and

$$\boldsymbol{H} = \boldsymbol{B} - 4\pi\boldsymbol{M}\,, \tag{24.8}$$

if we recall that $\varrho = \varrho_t - \operatorname{div}\boldsymbol{P}$, as was shown in Section 7. This results in

$$\operatorname{curl}\boldsymbol{H} = \frac{4\pi}{c}\,\boldsymbol{i}_c + \frac{1}{c}\,\dot{\boldsymbol{D}}\,, \tag{24.2''}$$

$$\operatorname{div}\boldsymbol{D} = 4\pi\varrho_t\,. \tag{24.4''}$$

Just as Eq. [24.5] follows from Eqs. [24.2] and [24.4], the continuity equation for true (conduction) charges,

$$\dot{\varrho}_t + \operatorname{div}\boldsymbol{i}_c = 0\,, \tag{24.5'}$$

follows from Eqs. [24.2″] and [24.4″].

To the above equations must be added the phenomenological (semiempirical) *equations*:

$$\boldsymbol{D} = \varepsilon\boldsymbol{E}\,, \tag{24.9}$$

$$\boldsymbol{B} = \mu\boldsymbol{H}\,, \tag{24.10}$$

$$\boldsymbol{i}_c = \sigma\boldsymbol{E}\,. \tag{24.11}$$

The quantities ε, μ, and σ are not really constants; they are different for slowly and rapidly varying fields. For periodic fields

$$\boldsymbol{F} = \boldsymbol{F}_0(\boldsymbol{x})\,e^{i\omega t}\,,$$

where $\boldsymbol{F}$ represents any of the quantities $\boldsymbol{D}$, $\boldsymbol{E}$, $\boldsymbol{B}$, $\boldsymbol{H}$, or $\boldsymbol{i}$, they are functions of the frequency:

$$\varepsilon(\omega)\,,\qquad \mu(\omega)\,,\qquad \sigma(\omega)\,.$$

In the limit $\omega \to 0$, they converge to the ε, μ, and σ for static fields. The theoretical determination of the frequency dependence must be made with the aid of special conceptual models for the medium. In the final analysis, quantum theory must be invoked since the question of atomic structure is involved. For arbitrary (nonperiodic) fields, ε, μ, and σ are not even numbers but rather operators. The phenomenological equations [24.9], [24.10], and [24.11] are less general than the other equations.

One consequence of the continuity equation [24.5′] and the phenomenological equations [24.9] and [24.11] is the following:

$$\operatorname{div} \boldsymbol{i}_c = \sigma \operatorname{div} \boldsymbol{E} = \frac{\sigma}{\varepsilon} \operatorname{div} \boldsymbol{D} = \frac{4\pi\sigma}{\varepsilon} \varrho_t ,$$

$$\dot{\varrho}_t + \frac{4\pi\sigma}{\varepsilon} \varrho_t = 0 .$$

Upon integrating,

$$\varrho_t = \varrho_t^0 e^{-\frac{4\pi\sigma}{\varepsilon} t} .$$

Thus, a finite time is required for the true charge density in a conductor to disappear. This is called the *relaxation time*.

25. ELECTROMAGNETIC WAVES IN VACUUM

The existence of electromagnetic waves is a fundamental consequence of the Maxwell equations. In vacuum,

$$\boldsymbol{i}_c = 0 , \qquad \varrho_t = 0 .$$

From [24.1] it follows, by taking the divergence, that

$$\frac{\partial}{\partial t} \operatorname{div} \boldsymbol{H} = 0 .$$

Equation [24.3] is, therefore, not independent of Eq. [24.1], but it does, however, assert somewhat more. For periodic fields Eq. [24.3] is, indeed, even a consequence of Eq. [24.1].

Analogously, Eq. [24.2], with $\boldsymbol{i}=0$, leads to

$$\frac{\partial}{\partial t}\operatorname{div}\boldsymbol{E}=0\,,$$

and, for periodic fields, even to

$$\operatorname{div}\boldsymbol{E}=0\,, \qquad [25.1]$$

this latter result following in all generality from Eq. [24.4] when $\varrho=0$.

The additional term $\dot{\boldsymbol{E}}/c$ in Eq. [24.2] has the physical consequence that electromagnetic waves can exist. That is, upon eliminating $\boldsymbol{H}$ from Eqs. [24.1] and

$$\operatorname{curl}\boldsymbol{H}=\frac{1}{c}\dot{\boldsymbol{E}} \qquad [25.2]$$

by forming

$$\operatorname{curl}\operatorname{curl}\boldsymbol{E}=\operatorname{grad}\operatorname{div}\boldsymbol{E}-\nabla^2\boldsymbol{E}=-\frac{1}{c^2}\ddot{\boldsymbol{E}},$$

one obtains, because of Eq. [25.1], the famous *wave equation* for $\boldsymbol{E}$:

$$\nabla^2\boldsymbol{E}-\frac{1}{c^2}\ddot{\boldsymbol{E}}=0\,. \qquad [25.3]$$

Equations [24.1], [25.2], [24.3] and [25.1] remain correct upon the substitution

$$\boldsymbol{E}\to\boldsymbol{H}\,, \qquad \boldsymbol{H}\to-\boldsymbol{E}\,. \qquad [25.4]$$

Thus, it also is true that

$$\nabla^2\boldsymbol{H}-\frac{1}{c^2}\ddot{\boldsymbol{H}}=0\,. \qquad [25.5]$$

As an example of a solution of the wave equations [25.3] and [25.5], we consider *plane waves*:

$$\boldsymbol{E}=\boldsymbol{E}_0e^{i(\boldsymbol{k}\cdot\boldsymbol{x}-\omega t)}\,, \qquad \boldsymbol{H}=\boldsymbol{H}_0e^{i(\boldsymbol{k}\cdot\boldsymbol{x}-\omega t)}\,. \qquad [25.6]$$

(In these equations, the real parts should actually be considered since only they have physical significance.) We have set

$$\omega = 2\pi\nu\,, \tag{25.7}$$

$$\boldsymbol{k} = k\boldsymbol{n}\,, \tag{25.8}$$

where ν is the frequency and $\boldsymbol{n}$ is the wave normal; that is, the unit vector $\boldsymbol{n}$ is perpendicular to the surfaces of constant phase, the so-called wave surfaces: $(\boldsymbol{n}\cdot\boldsymbol{x}) =$ constant. The phase also remains unchanged at a point a distance $\lambda = 2\pi/k$ away along the $\boldsymbol{n}$ direction. Thus,

$$k = \frac{2\pi}{\lambda}\,, \tag{25.9}$$

where λ represents the wavelength. For solutions of the form given in [25.6], the differential operators are equivalent to certain multiplications:

$$\begin{aligned} &\frac{\partial}{\partial \boldsymbol{x}} \sim i\boldsymbol{k}\,, \qquad &&\frac{\partial}{\partial t} \sim -i\omega\,, \\ &\nabla^2 \sim -k^2\,, \qquad &&\frac{\partial^2}{\partial t^2} \sim -\omega^2\,. \end{aligned} \tag{25.10}$$

These substituted into Eq. [25.3] yield

$$\left(-k^2 + \frac{1}{c^2}\,\omega^2\right) E = 0\,, \qquad k^2 = \frac{\omega^2}{c^2}\,. \tag{25.11}$$

In order to see a constant phase, one must advance with the velocity ω/k in the $\boldsymbol{n}$ direction, since then

$$x_{\parallel} = \frac{\omega}{k}\,t\,, \qquad k(\boldsymbol{n}\cdot\boldsymbol{x}) = \omega t\,.$$

Thus,

$$\frac{\omega}{k} = c \tag{25.12}$$

is the ***phase velocity***. The quantity c is contained in the differential equations and is independent of ω. Furthermore, it follows from Eqs. [24.3] and [25.1] that

$$i(\boldsymbol{k}\cdot\boldsymbol{H}) = 0\,, \qquad i(\boldsymbol{k}\cdot\boldsymbol{E}) = 0\,,$$

so that

$$\boldsymbol{n}\cdot\boldsymbol{H} = \boldsymbol{n}\cdot\boldsymbol{E} = 0\,.$$

Thus both $\boldsymbol{H}$ and $\boldsymbol{E}$ are perpendicular to $\boldsymbol{n}$; that is, *electromagnetic waves are transverse.*

Because of Eqs. [25.10], it follows from Eq. [25.2] that

$$i(\boldsymbol{k}\times\boldsymbol{H}) = -\frac{i\omega}{c}\boldsymbol{E}\,,$$

while from Eqs. [25.8] and [25.12] there results

$$\boldsymbol{n}\times\boldsymbol{H} = -\boldsymbol{E} \qquad \text{or} \qquad \boldsymbol{E} = \boldsymbol{H}\times\boldsymbol{n}\,.$$

Thus $\boldsymbol{E}$ is perpendicular to $\boldsymbol{n}$ and to $\boldsymbol{H}$. Hence, $\boldsymbol{E}\cdot\boldsymbol{H}=0$. Furthermore,

$$|\boldsymbol{E}| = |\boldsymbol{H}| \qquad \text{or} \qquad E^2 = H^2\,, \qquad [25\ 13]$$

since $\boldsymbol{n}$ is a unit vector and $\boldsymbol{H}$ is perpendicular to $\boldsymbol{n}$. Since the equations remain correct under the substitution given in Eq. [25.4], we have

$$\boldsymbol{H} = -(\boldsymbol{E}\times\boldsymbol{n}) = \boldsymbol{n}\times\boldsymbol{E}\,,$$

and it is seen that ***E***, ***H*** *and* ***n***, *in that order, form a right-handed orthogonal system.*

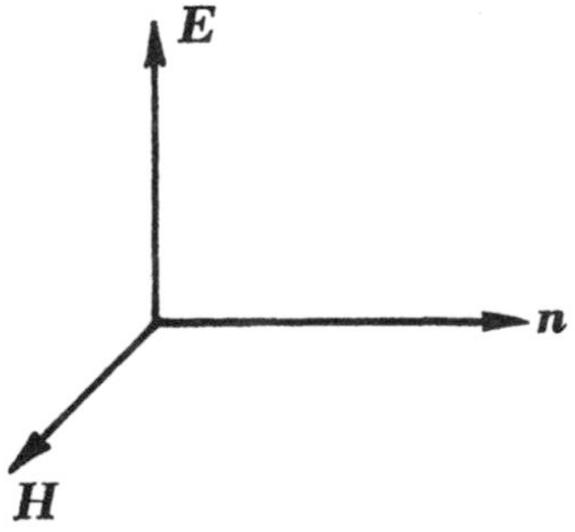

Figure 25.1

There follows the relation

$$\boldsymbol{E}\times\boldsymbol{H} = \boldsymbol{n}E^2 = \boldsymbol{n}H^2 = \tfrac{1}{2}(E^2+H^2)\boldsymbol{n}\,, \qquad [25.14]$$

which we shall use later.

These results obtained by Maxwell refer to electromagnetic waves. It is natural to consider light as being waves of this kind since

1. the transversality, which is very difficult to explain in terms of a mechanistic aether concept, results quite naturally, and
2. it turns out that c, which was defined as a pure electromagnetic quantity and, as such, is determined by the magnetic field due to a current, agrees with the empirically determined velocity of light.

Light, therefore, simply signifies electromagnetic waves of certain special wavelengths (*electromagnetic theory of light*). Hertz and others later discovered electromagnetic waves of other wavelengths (radio waves, infrared).

26. CONSERVATION OF ENERGY AND MOMENTUM

We start with the Maxwell equations, [24.1], [24.2], [24.3], and [24.4].

According to Eqs. [17.5], the force per unit volume is

$$\boldsymbol{k} = \varrho\boldsymbol{E} + \frac{1}{c}\,\boldsymbol{i}\times\boldsymbol{H}\,. \qquad [26.1]$$

Thus, the work done per unit volume, per unit time, is

$$A = \boldsymbol{i}\cdot\boldsymbol{E}\,. \qquad [26.2]$$

a. Conservation of energy

Multiplying Eq. [24.2] by $c\boldsymbol{E}/4\pi$ yields

$$A + \frac{1}{4\pi}\,\boldsymbol{E}\cdot\dot{\boldsymbol{E}} - \frac{c}{4\pi}\,\boldsymbol{E}\cdot\operatorname{curl}\boldsymbol{H} = 0\,,$$

while Eq. [24.1] multiplied by $c\boldsymbol{H}/4\pi$ gives

$$\frac{1}{4\pi}\boldsymbol{H}\cdot\dot{\boldsymbol{H}}+\frac{c}{4\pi}\boldsymbol{H}\cdot\operatorname{curl}\boldsymbol{E}=0\,.$$

Adding these two equations and using the fact that

$$\operatorname{div}\boldsymbol{E}\times\boldsymbol{H}=\boldsymbol{H}\cdot\operatorname{curl}\boldsymbol{E}-\boldsymbol{E}\cdot\operatorname{curl}\boldsymbol{H}$$

results in the relation

$$A+\frac{\partial W}{\partial t}+\operatorname{div}\boldsymbol{S}=0\,. \qquad [26.3]$$

Here,

$$W=\frac{1}{8\pi}(E^2+H^2) \qquad [26.4]$$

is the *energy density* and

$$\boldsymbol{S}=\frac{c}{4\pi}\boldsymbol{E}\times\boldsymbol{H} \qquad [26.5]$$

is the *Poynting vector*. Equation [26.3] is the law of conservation of energy as it applies to field theory. Upon integration, there results

$$\int A\,\mathrm{d}V+\frac{\mathrm{d}}{\mathrm{d}t}\int W\,\mathrm{d}V+\oint S_n\,\mathrm{d}f=0\,. \qquad [26.6]$$

Here, the first term is the total work done per second by the system, and the second term the energy increase of the system per second. Thus, the third term clearly must represent the total energy flowing out of the bounding surface of the system per second. Therefore, $\boldsymbol{S}$ has the significance of an energy current density.

If the bounding surfaces are absorptive or else are so far away that the fields have not yet reached them, then the third integral vanishes. The work done by the system is then equal to its increase in energy. This is always the case for static fields since then $|\boldsymbol{S}|$ at large distances falls off at least as fast as $1/r^4$.

For plane waves, according to Eq. [25.14],

$$\boldsymbol{S} = nc W . \qquad [26.7]$$

This is very satisfying since it implies that the velocity with which the energy propagates is equal to the phase velocity of the waves.

b. Conservation of momentum

This consideration is similar to that of Section 18, although here there are additional terms.

From Eq. [26.1] and the Maxwell equations [24.1] to [24.4], it follows that

$$4\pi \boldsymbol{K} = 4\pi \int \boldsymbol{k} \, \mathrm{d}V = 4\pi \int \left(\varrho \boldsymbol{E} + \frac{1}{c} \boldsymbol{i} \times \boldsymbol{H}\right) \mathrm{d}V$$
$$= \int \left\{\boldsymbol{E} \cdot \operatorname{div} \boldsymbol{E} + \operatorname{curl} \boldsymbol{H} \times \boldsymbol{H} - \frac{1}{c} \dot{\boldsymbol{E}} \times \boldsymbol{H}\right\} \mathrm{d}V .$$

Because

$$\boldsymbol{H} \cdot \operatorname{div} \boldsymbol{H} = 0 \qquad \text{and} \qquad \left(\operatorname{curl} \boldsymbol{E} + \frac{1}{c} \dot{\boldsymbol{H}}\right) \times \boldsymbol{E} = 0 ,$$

there results

$$0 = \int \left\{\boldsymbol{H} \cdot \operatorname{div} \boldsymbol{H} + \operatorname{curl} \boldsymbol{E} \times \boldsymbol{E} - \frac{1}{c} \boldsymbol{E} \times \dot{\boldsymbol{H}}\right\} \mathrm{d}V .$$

With the abbreviations

$$\boldsymbol{J}_e = \int \{\boldsymbol{E} \cdot \operatorname{div} \boldsymbol{E} + \operatorname{curl} \boldsymbol{E} \times \boldsymbol{E}\} \mathrm{d}V ,$$

$$\boldsymbol{J}_m = \int \{\boldsymbol{H} \cdot \operatorname{div} \boldsymbol{H} + \operatorname{curl} \boldsymbol{H} \times \boldsymbol{H}\} \mathrm{d}V ,$$

we then have

$$4\pi \boldsymbol{K} = \boldsymbol{J}_e + \boldsymbol{J}_m - \frac{1}{c} \frac{\mathrm{d}}{\mathrm{d}t} \int \boldsymbol{E} \times \boldsymbol{H} \, \mathrm{d}V .$$

Now, it has been shown in Section 18 that

$$\frac{1}{4\pi} (\boldsymbol{J}_e + \boldsymbol{J}_m) = \oint \boldsymbol{T}_n \, \mathrm{d}f ,$$

where

$$T_{ik} = T_{ik}^{(e)} + T_{ik}^{(m)} \qquad [26.8]$$

is given in Eq. [18.7]. Thus, we obtain

$$\boldsymbol{K} = \oint \boldsymbol{T}_n \,\mathrm{d}f - \frac{1}{c^2}\frac{\mathrm{d}}{\mathrm{d}t}\int \boldsymbol{S}\,\mathrm{d}V\,. \qquad [26.9]$$

The second term on the right is exactly zero for steady-state fields and is negligible for quasi-static fields. The differential form of this relation is

$$k_i = \sum_k \frac{\partial T_{ik}}{\partial x_k} - \frac{1}{c^2}\frac{\partial S_i}{\partial t} \qquad (i = 1, 2, 3)\,. \qquad [26.10]$$

The significance of Eqs. [26.9] and [26.10] is that ***momentum can be conserved only if a momentum as well as an energy is attributed to the electromagnetic field.*** If the total mechanical momentum is denoted by $\boldsymbol{P}_m$, then

$$\frac{\mathrm{d}\boldsymbol{P}_m}{\mathrm{d}t} = \boldsymbol{K}\,. \qquad [26.11]$$

From this it follows that an electromagnetic momentum

$$\boldsymbol{P}_{\mathrm{em}} = \frac{1}{c^2}\int \boldsymbol{S}\,\mathrm{d}V \qquad [26.12]$$

must be ascribed to the field. This implies a momentum density of

$$\boldsymbol{g} = \frac{1}{c^2}\boldsymbol{S} \qquad [26.13]$$

per cubic centimeter. If the bounding surfaces of the region of integration in Eq. [26.9] are so located that at some definite time they have not yet been reached by the fields, then

$$\oint \boldsymbol{T}_n \,\mathrm{d}f = 0\,,$$

$$\frac{\mathrm{d}}{\mathrm{d}t}(\boldsymbol{P}_m + \boldsymbol{P}_{\mathrm{em}}) = 0\,, \qquad \boldsymbol{P}_m + \boldsymbol{P}_{\mathrm{em}} = \text{constant}\,, \qquad [26.14]$$

as shown by Poincaré and Einstein. In general, however, $\boldsymbol{P}_m$ and $\boldsymbol{P}_{\mathrm{em}}$ are not separately constant.

c. *Radiation pressure*

A plane wave is a mathematical idealization: it is infinite in extent and has, therefore, infinite energy; its frequency is sharply defined. We can, however, allow the wavelength to vary within some arbitrarily small interval. This is sufficient to produce a finite wave train (theory of the Fourier integral). Such a wave train having very many humps but,

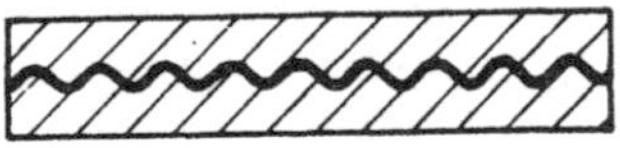

Figure 26.1

nevertheless, being limited in extent is called a *wave packet* (Fig. 26.1). The energy and momentum are then given by

$$E = \int W \, \mathrm{d}V , \qquad \boldsymbol{P}_{\mathrm{em}} = \frac{1}{c^2} \int \boldsymbol{S} \, \mathrm{d}V = \boldsymbol{n} \frac{E}{c} ,$$

according to Eq. [26.7]. Thus, a wave train whose dimensions are large as compared with its wavelength, propagating in a definite direction, has the momentum

$$\frac{\text{energy}}{\text{velocity of light}} .$$

This suffices for calculating the radiation pressure.[1]

Simplest case. Total reflection at normal incidence (Fig. 26.2). The relation given by Eq. [26.9] is applied to

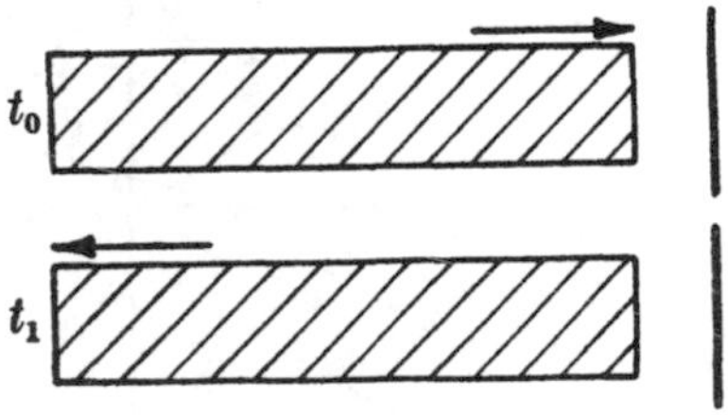

Figure 26.2

[1] The radiation pressure can also be calculated directly from Eq. [26.1].

a surface at which the fields exist from a time t_0 until a time t_1. The electromagnetic momentum changes by an amount $2E/c$ and a reaction is clearly exerted on the surface. The change in mechanical momentum is

$$\Delta P_m = 2\frac{E}{c}.$$

If T denotes the length of time during which the wave train is in contact with the surface, then to sufficient accuracy,

$$E = TFI,$$

where F is the exposed area and $I = \bar{S}_x$ is the intensity of the radiation. Thus,

$$\Delta P_m = pFT,$$

where

$$p = 2\frac{I}{c} \qquad [26.15]$$

is the radiation pressure.

It is easily seen that *for a wave packet incident at an angle* ϑ,

$$p = 2\frac{I}{c}\cos^2\vartheta. \qquad [26.16]$$

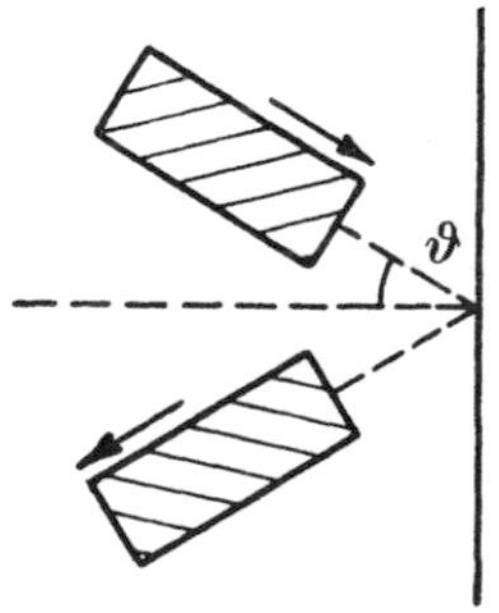

Figure 26.3

For *total absorption*,

$$\Delta P_m = \frac{E}{c},$$

$$p = \frac{I}{c}. \qquad [26.17]$$

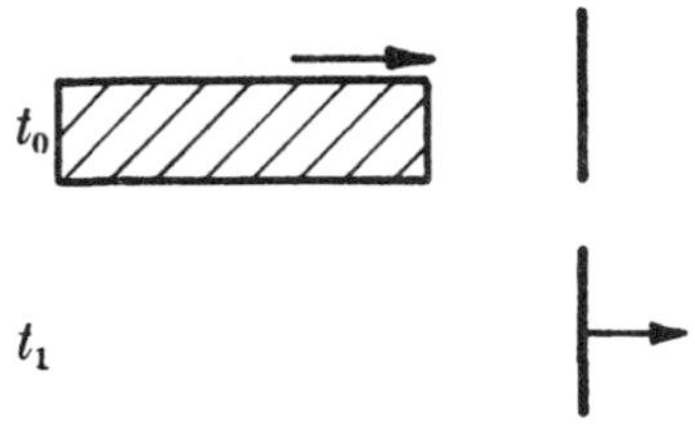

Figure 26.4

These results cannot be obtained without the electromagnetic theory of light. The essential fact that we have used is that the electromagnetic momentum per unit volume is $\boldsymbol{S}/c^2$. Radiation pressure was first experimentally detected by Lebedev.

This association of momentum with the electromagnetic field is quite natural. Without it, conservation of energy and momentum at each instant of time cannot be obtained. If light is emitted and then absorbed, the conservation law for mechanical momentum is, indeed, once again valid after the absorption; however, it seems unnatural to eliminate the field since it is not apparent why more reality should be ascribed to the material particles than to the field.

Thus, on the one hand, the aether (by aether we now mean all possible electromagnetic fields) is described independently without reference to mechanical quantities, and it is pointless to speak about a motion of the aether. On the other hand, the equations for a perfect vacuum ($\varrho = 0$, $\boldsymbol{i} = 0$) are only an idealization, since electromagnetic fields can be produced and detected only with the use of the mechanically describable particles that carry charge. A noteworthy duality is encountered here [A-1].

27. ELECTROMAGNETIC WAVES IN MATERIAL MEDIA

The differences between material media and the vacuum are attributed to magnetization and polarization. If we allow ε and μ to differ from unity, then the Maxwell equations of [24.1′], [24.2′], [24.3′] and [24.4′] become, in the absence of true charges or currents ($\varrho_t = \boldsymbol{i}_c = 0$),

$$\operatorname{curl} \boldsymbol{E} = -\frac{\mu}{c}\dot{\boldsymbol{H}}, \qquad \operatorname{div} \boldsymbol{H} = 0,$$

$$\operatorname{curl} \boldsymbol{H} = +\frac{\varepsilon}{c}\dot{\boldsymbol{E}}, \qquad \operatorname{div} \boldsymbol{E} = 0.$$

The phase velocity of the waves is then

$$v = \frac{c}{\sqrt{\varepsilon\mu}}. \qquad [27.1]$$

For the index of refraction, defined by

$$v \equiv \frac{c}{n}, \qquad [27.2]$$

we obtain

$$n^2 = \varepsilon\mu. \qquad [27.3]$$

For plane waves, in analogy to Section 25, it is found that $\boldsymbol{E}$, $\boldsymbol{H}$, *and* $\boldsymbol{n}$ *again form a right-handed orthogonal system.* However, the ratio of the $\boldsymbol{E}$ and $\boldsymbol{H}$ amplitudes is no longer unity. Instead,

$$i(\boldsymbol{k} \times \boldsymbol{E}) = ik(\boldsymbol{n} \times \boldsymbol{E}) = +\frac{\mu}{c} i\omega \boldsymbol{H},$$

$$k|\boldsymbol{E}| = \mu\frac{\omega}{c}|\boldsymbol{H}|,$$

$$k^2 = \frac{\varepsilon\mu}{c^2}\omega^2, \qquad k = \frac{\omega}{c}\sqrt{\varepsilon\mu}, \qquad [27.4]$$

$$|\boldsymbol{H}| = \sqrt{\frac{\varepsilon}{\mu}}\,|\boldsymbol{E}|.$$

The last equation can be remembered most easily in the form

$$\varepsilon E^2 = \mu H^2. \qquad [27.5]$$

With respect to Eq. [27.3], it may be remarked that para- and diamagnetism are always very weak, so that for non-ferromagnetic substances it is essentially true that $\mu = 1$, $n^2 = \varepsilon(\omega)$, where $\lim_{\omega \to 0} n^2(\omega) = \varepsilon_{stat}$ (static dielectric constant). This can be explained in terms of electron theory. For ferromagnetic materials, μ is also essentially equal to 1 in the optical region of ω since ferromagnetism does not occur at high frequencies.

28. RADIATION OF ELECTROMAGNETIC WAVES

In this section the generalization of the fields produced by static and quasi-static dipoles is described.

a. The electromagnetic potentials. The inhomogeneous wave equation

We start from the Maxwell equations in vacuum, [24.1], [24.2], [24.3], and [24.4]. From Sections 20 and 21 we already know that Eqs. [24.1] and [24.3] are identically satisfied by the introduction of the potentials $\boldsymbol{A}$ and φ:

$$\boldsymbol{H} = \operatorname{curl} \boldsymbol{A}, \qquad [28.1]$$

$$\boldsymbol{E} = -\frac{1}{c}\dot{\boldsymbol{A}} - \operatorname{grad}\varphi. \qquad [28.2]$$

Introducing $\boldsymbol{A}$ and φ into Eq. [24.2] yields

$$\operatorname{grad}\left(\operatorname{div}\boldsymbol{A} + \frac{1}{c}\dot{\varphi}\right) - \nabla^2\boldsymbol{A} + \frac{1}{c^2}\ddot{\boldsymbol{A}} = \frac{4\pi}{c}\boldsymbol{i}$$

because of Eq. [15.9]. Putting $\boldsymbol{A}$ and φ into Eq. [24.4] gives

$$-\frac{1}{c}\operatorname{div}\dot{\boldsymbol{A}} - \nabla^2\varphi = 4\pi\varrho .$$

Here, it is useful to rewrite this equation slightly:

$$-\frac{1}{c}\frac{\partial}{\partial t}\left(\operatorname{div}\boldsymbol{A} + \frac{1}{c}\dot{\varphi}\right) - \nabla^2\varphi + \frac{1}{c^2}\ddot{\varphi} = 4\pi\varrho .$$

It is now seen that it is convenient to require that

$$\operatorname{div} \boldsymbol{A} + \frac{1}{c}\dot{\varphi} = 0 \qquad (\textit{the Lorentz condition}), \tag{28.3}$$

rather than to demand that $\operatorname{div} \boldsymbol{A} = 0$ as was done for the quasi-static case. Thus, one obtains the ***inhomogeneous wave equations*** for φ and $\boldsymbol{A}$:

$$-\nabla^2\varphi + \frac{1}{c^2}\ddot{\varphi} = 4\pi\varrho\,, \tag{28.4}$$

$$-\nabla^2\boldsymbol{A} + \frac{1}{c^2}\ddot{\boldsymbol{A}} = \frac{4\pi}{c}\boldsymbol{i}\,. \tag{28.5}$$

The three equations [28.3], [28.4], and [28.5] are consistent, since

$$\dot{\varrho} + \operatorname{div} \boldsymbol{i} = 0\,. \tag{28.6}$$

In vacuum (where $\varrho = 0, \boldsymbol{i} = 0$) these relations reduce to the homogeneous wave equations.

The integration of Eqs. [28.3], [28.4], and [28.5] will be carried out in two different ways.

b. Integration of the inhomogeneous wave equations. First method

First of all, we assume that *the fields are periodic in time.* This results in an important simplification. Let the charge and current densities be

$$\varrho = \varrho_0 e^{-i\omega t}, \qquad \boldsymbol{i} = \boldsymbol{i}_0 e^{-i\omega t}. \tag{28.7}$$

Then,

$$\varphi = \varphi_0 e^{-i\omega t}, \qquad \boldsymbol{A} = \boldsymbol{A}_0 e^{-i\omega t},$$

from which it follows that

$$\frac{\partial}{\partial t} \sim -i\omega\,, \qquad \frac{\partial^2}{\partial t^2} \sim -\omega^2\,.$$

With

$$\frac{\omega}{c} = k\,, \tag{28.8}$$

the differential equations [28.3], [28.4], and [28.5] yield

$$\operatorname{div} \boldsymbol{A} - ik\varphi = 0\,, \qquad [28.3']$$

$$\nabla^2\varphi + k^2\varphi = -4\pi\varrho\,, \qquad [28.4']$$

$$\nabla^2\boldsymbol{A} + k^2\boldsymbol{A} = -\frac{4\pi}{c}\,\boldsymbol{i}\,. \qquad [28.5']$$

These equations are easily integrated by the use of a method based on one employed in Section 6. There it is shown that the solution of

$$-\nabla^2\varphi = 4\pi\varrho$$

is

$$\varphi_P = \int \frac{\varrho_Q \cdot \mathrm{d}V_Q}{r_{PQ}}\,.$$

To obtain this result, Green's theorem,

$$\int (\varphi\,\nabla^2\psi - \psi\,\nabla^2\varphi)\,\mathrm{d}V = \oint \left(\varphi\,\frac{\partial\psi}{\partial n} - \psi\,\frac{\partial\varphi}{\partial n}\right)\mathrm{d}f\,,$$

is used with $\psi = 1/r_{PQ}$ so that

$$\nabla^2\psi = 0 \quad \text{for} \quad P \neq Q\,.$$

For the present case one must set

$$\psi = \frac{e^{ikr_{PQ}}}{r_{PQ}} \qquad [28.9]$$

in Green's theorem. This ψ satisfies the equation

$$\nabla^2\psi + k^2\psi = 0 \quad \text{for} \quad r \neq 0\,. \qquad [28.10]$$

Proof: In polar coordinates Eq. [28.10] becomes

$$\frac{1}{r}\left\{\frac{\mathrm{d}^2(r\psi)}{\mathrm{d}r^2} + k^2(r\psi)\right\} = 0\,,$$

which has the solution

$$r\psi = e^{\pm ikr}\,.$$

(The reason that we have taken the $+i$ rather than the $-i$ solution in Eq. [28.9] will be made clear later.)

For the region of integration V, we again take the interior of a large sphere K about the point P. The point P itself is excluded by means of a small sphere k. Thus, because of [28.10] and [28.4′],

$$\varphi \nabla^2\psi - \psi \nabla^2\varphi = \varphi(\nabla^2\psi + k^2\psi) - \psi(\nabla^2\varphi + k^2\varphi) = +4\pi\varrho \frac{e^{ikr_{PQ}}}{r_{PQ}} .$$

This, substituted into Green's theorem, gives

$$4\pi \int_V \varrho_Q \frac{e^{ikr_{PQ}}}{r_{PQ}} \mathrm{d}V_Q = \oint_k \left(\varphi \frac{\partial}{\partial n}\frac{e^{ikr}}{r} - \frac{e^{ikr}}{r}\frac{\partial\varphi}{\partial n}\right) \mathrm{d}f$$
$$+ \oint_K \left(\varphi \frac{\partial}{\partial n}\frac{e^{ikr}}{r} - \frac{e^{ikr}}{r}\frac{\partial\varphi}{\partial n}\right) \mathrm{d}f .$$

Now, at the surface of the small sphere,

$$\partial/\partial n = -\partial/\partial r, \qquad \mathrm{d}f = r^2\,\mathrm{d}\Omega ,$$

where $\mathrm{d}\Omega$ is the solid angle. Also,

$$\frac{\partial}{\partial r}\frac{e^{ikr}}{r} = \frac{1}{r^2}(ikr - 1)\, e^{ikr} .$$

Thus, the integral over the small sphere becomes

$$\oint_k \mathrm{d}\Omega \left\{\varphi_Q - ikr\varphi_Q + r\frac{\partial\varphi_Q}{\partial r}\right\} e^{ikr} .$$

Because $\partial/\partial n = \partial/\partial R$, the integral over the large sphere becomes

$$-\int_K \mathrm{d}\Omega \left\{\varphi + R\left(\frac{\partial\varphi}{\partial R} - ik\varphi\right)\right\} e^{ikR} .$$

We now consider the limit in which the small sphere vanishes, so that $r \to 0$. Hence,

$$\int_k \to 4\pi\varphi_P .$$

As the large sphere becomes infinite, $R \to \infty$ and $\lim_{R\to\infty} \varphi = 0$. In addition, we require that

$$\lim_{R\to\infty} R\left(\frac{\partial\varphi}{\partial R} - ik\varphi\right) = 0 . \qquad [28.11]$$

This is satisfied if one demands that in the asymptotic region, for large R,

$$\varphi \sim \frac{e^{ikR}}{R} .$$

Since

$$\left.\begin{array}{ll} e^{i(kr+\omega t)} & \text{describes outgoing waves} \\ e^{i(kr-\omega t)} & \text{describes incoming waves} \end{array}\right\},$$

the significance of Eq. [28.11] is that *only outgoing waves* are considered. (Had $-ik$ been taken in the previous equations instead of $+ik$, then *incoming* waves would have been obtained.) Equation [28.11] is called the *radiation condition* (after Sommerfeld). This condition does not follow from the differential equations but is, instead, an important additional physical requirement. This condition must be imposed in order that the solution be unique.

The integral over the large sphere now approaches zero as $R \to \infty$ and there results

$$\varphi_P = \int \varrho_Q \frac{e^{ikr_{PQ}}}{r_{PQ}} \, dV_Q . \qquad [28.12]$$

Using an analogous procedure for $\boldsymbol{A}$, one obtains

$$\boldsymbol{A}_P = \frac{1}{c}\int \boldsymbol{i}_Q \frac{e^{ikr_{PQ}}}{r_{PQ}} \, dV_Q . \qquad [28.13]$$

The subsidiary Lorentz condition of Eq. [28.3′] is also satisfied because of the continuity equation [28.6] which, in our case, can be written as

$$\frac{1}{c} \operatorname{div} \boldsymbol{i} - ik\varrho = 0 .$$

Employing the identity

$$\operatorname{div}(\boldsymbol{a}f) \equiv f \operatorname{div} \boldsymbol{a} + \boldsymbol{a} \cdot \operatorname{grad} f$$

twice, there results

$$\operatorname{div} \boldsymbol{A} = \frac{1}{c}\int dV_Q \boldsymbol{i}_Q \cdot \operatorname{grad}_P \frac{e^{ikr_{PQ}}}{r_{PQ}} = -\frac{1}{c}\int dV_Q \boldsymbol{i}_Q \cdot \operatorname{grad}_Q \frac{e^{ikr_{PQ}}}{r_{PQ}}$$

$$= -\frac{1}{c}\int dV_Q \operatorname{div}_Q \left(\boldsymbol{i}_Q \frac{e^{ikr_{PQ}}}{r_{PQ}}\right) + \frac{1}{c}\int dV_Q (\operatorname{div}_Q \boldsymbol{i}_Q) \frac{e^{ikr_{PQ}}}{r_{PQ}} .$$

The first integral vanishes when transformed into a surface integral by Gauss's theorem and evaluated over the infinite sphere. Thus,

$$\operatorname{div} \boldsymbol{A} = i \cdot k \int dV_Q \, \varrho_Q \frac{e^{ikr_{PQ}}}{r_{PQ}} = ik\varphi ,$$

which is the Lorentz condition for periodic fields.

Equation [28.3′] is, in addition, very useful for determining φ when $\boldsymbol{A}$ is known. If $t \to t - r/c$ is substituted into Eq. [28.7], then, because of Eq. [28.8], we have the transformation

$$e^{-i\omega t} \to e^{-i\omega t + ikr} .$$

The solutions of Eqs. [28.12] and [28.13] then can be expressed as

$$\varphi_P(t) = \int \varrho_Q \left(t - \frac{r_{PQ}}{c}\right) \frac{dV_Q}{r_{PQ}} , \qquad [28.14]$$

$$\boldsymbol{A}_P(t) = \frac{1}{c}\int \boldsymbol{i}_Q \left(t - \frac{r_{PQ}}{c}\right) \frac{dV_Q}{r_{PQ}} . \qquad [28.15]$$

This form of the solution is much more general than that of Eqs. [28.12] *and* [28.13]. It is not tied to a periodic ($\sim e^{-i\omega t}$) time dependence. That is, if ϱ and $\boldsymbol{i}$ are linear superpositions of periodic quantities (Fourier integrals),

$$\varrho(t)=\int_{-\infty}^{+\infty}\varrho(\omega)\,e^{-i\omega t}\,d\omega\,,$$

$$\boldsymbol{i}(t)=\int_{-\infty}^{+\infty}\boldsymbol{i}(\omega)\,e^{-i\omega t}\,d\omega\,,$$

then the corresponding solution is also a linear superposition:

$$\varphi(t)=\int_{-\infty}^{+\infty}\varphi(\omega)\;e^{-i\omega t}\,d\omega\,,$$

$$\boldsymbol{A}(t)=\int_{-\infty}^{+\infty}\boldsymbol{A}(\omega)\,e^{-i\omega t}\,d\omega\,.$$

That is, *Eqs.* [28.14] *and* [28.15] *are valid as long as* ϱ *and* $\boldsymbol{i}$ *can be represented by a Fourier expansion in time.* In the form given by Eqs. [28.14] and [28.15], φ_P and $\boldsymbol{A}_P$ are called ***retarded potentials*** because of the appearance of $t-r/c$. They correspond to the radiation condition. The *advanced potentials* would be

$$\varphi_P(t)=\int\varrho_Q\left(t+\frac{r_{PQ}}{c}\right)\frac{dV_Q}{r_{PQ}}\,,$$

$$\boldsymbol{A}_P(t)=\frac{1}{c}\int\boldsymbol{i}_Q\left(t+\frac{r_{PQ}}{c}\right)\frac{dV_Q}{r_{PQ}}\,.$$

These potentials represent incoming spherical waves. They are solutions of the same differential equations but are, however, not easily realizable in nature. In an infinite space, nature favors the first set of solutions.

c. *Discussion of the formulas for the periodic case*

A general consequence of the solutions of Eqs. [28.12] and [28.13] is the fact that a *radiation zone* is formed at great distances from the current system (i.e., the region where ϱ and $\boldsymbol{i}$ are significantly different from zero). If the charges are restricted to a finite region of space, the static and quasi-static fields fall off as $1/r^2$. For the case of rapidly varying fields, not only the potentials but also the field intensities fall off as $1/r$.

For the following considerations it is important to differentiate between three different lengths:

d = the linear extention of the current system;

R = the distance from the field point P to the current system;

$\lambda = 2\pi/k$ = the wavelength.

1. *General discussion of the radiation zone.* We now consider a field point P such that

$$R \gg d \qquad \text{and} \qquad R \gg \lambda$$

(that is, $kR \gg 1$). These two conditions characterize the

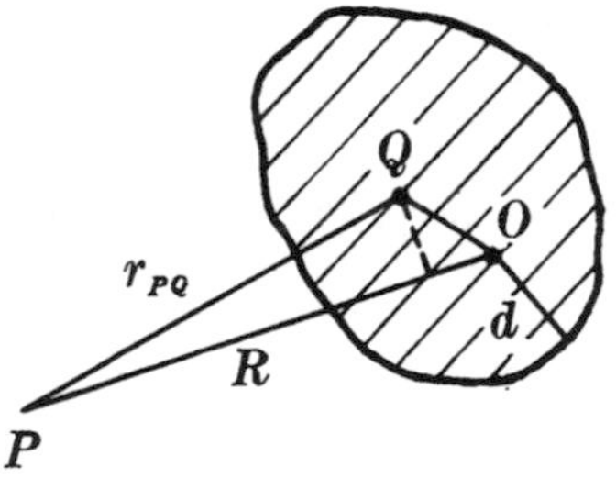

Figure 28.1

radiation zone. Nothing at all is presupposed about the ratio d/λ.

If a point O within the current system is chosen as the origin of coordinates, then

$$r_{PQ} = R - \boldsymbol{n}\cdot\boldsymbol{x}_Q \ ,$$

where $\overline{OP} = R$, $\boldsymbol{x}_Q$ = position vector of a source point Q, and $\boldsymbol{n} = \boldsymbol{x}_P/r$ is a unit vector in the P direction. In Eqs. [28.12] and [28.13] the denominator r_{PQ} can be replaced by R (this cannot be done in the exponent). Then,

$$\varphi_P = \frac{e^{ikR}}{R}\int \mathrm{d}V_Q\,\varrho_Q\,e^{-ik(\boldsymbol{n}\cdot\boldsymbol{x}_Q)}\,,$$

$$\boldsymbol{A}_P = \frac{e^{ikR}}{R}\,\frac{1}{c}\int \mathrm{d}V_Q\,\boldsymbol{i}_Q\,e^{-ik(\boldsymbol{n}\cdot\boldsymbol{x}_Q)}\,.$$

In the transition to the field intensities,

$$\boldsymbol{H} = \operatorname{curl}\boldsymbol{A}\,,\qquad \boldsymbol{E} = -\frac{1}{c}\dot{\boldsymbol{A}} - \operatorname{grad}\varphi\,,$$

the differentiation results in two terms: in the differentiation of $1/R$, the order of magnitude is decreased by a factor of $1/R$; in the differentiation of the exponential function, multiplication by k results. Since $kR \gg 1$, the first term can be neglected and we have

$$\frac{\partial}{\partial \boldsymbol{x}_P} \sim ik\frac{\partial R}{\partial \boldsymbol{x}_P} \sim ik\frac{\boldsymbol{x}_P}{R} \sim ik\boldsymbol{n}\,.$$

Thus,

$$\boldsymbol{H} = \operatorname{curl}\boldsymbol{A} = ik(\boldsymbol{n}\times\boldsymbol{A})$$

$$= \frac{ik}{c}\,\frac{e^{ikR}}{R}\int (\boldsymbol{n}\times\boldsymbol{i})_Q\,e^{ik(\boldsymbol{n}\cdot\boldsymbol{x}_Q)}\,\mathrm{d}V_Q\,. \qquad [28.16]$$

Because

$$\operatorname{div}\boldsymbol{A} = ik(\boldsymbol{n}\cdot\boldsymbol{A})$$

we obtain

$$\varphi = \boldsymbol{n}\cdot\boldsymbol{A}$$

from Eq. [28.3′]. Thus,

$$\boldsymbol{E} = ik\big(\boldsymbol{A} - \boldsymbol{n}(\boldsymbol{n}\cdot\boldsymbol{A})\big) = ik\boldsymbol{A}_\perp$$

$$= \frac{ik}{c}\,\frac{e^{ikR}}{R}\int (\boldsymbol{i}_\perp)_Q\,e^{-ik(\boldsymbol{n}\cdot\boldsymbol{x}_Q)}\,\mathrm{d}V_Q\,, \qquad [28.17]$$

where

$$\boldsymbol{i}_\perp = \boldsymbol{i} - \boldsymbol{n}(\boldsymbol{n}\cdot\boldsymbol{i})\,.$$

Since $\boldsymbol{n}\times\boldsymbol{i} = \boldsymbol{n}\times\boldsymbol{i}_\perp$, it is seen that

1. *only the components of* $\boldsymbol{i}$ *perpendicular to* $\boldsymbol{n}$ *are effective in producing the fields* **E** *and* **H**;

2. *the same relations hold between* **E**, **H**, *and* **n** *as for the case of plane waves* (Section 25):

$$\boldsymbol{H} = \boldsymbol{n} \times \boldsymbol{E}\,, \qquad \boldsymbol{H}\cdot\boldsymbol{n} = \boldsymbol{E}\cdot\boldsymbol{n} = \boldsymbol{E}\cdot\boldsymbol{H} = 0\,,$$

$$|\boldsymbol{E}|^2 = |\boldsymbol{H}|^2\,, \qquad \boldsymbol{S} = \frac{c}{4\pi}\boldsymbol{E}\times\boldsymbol{H} = \frac{c}{4\pi}|\boldsymbol{E}|^2\cdot\boldsymbol{n}\,.$$

Instead of Eq. [28.17] one can also write

$$\boldsymbol{E} = -\frac{1}{R}\frac{1}{c^2}\int\left(\frac{\partial \boldsymbol{i}_\perp}{\partial t}\right)_{\text{ret}} \mathrm{d}V\,. \qquad [28.18]$$

2. *Dipole radiation in the radiation zone.* In addition to the conditions $R \gg d$ and $R \gg \lambda$, we now also require that

$$\lambda \gg d$$

(that is, $kd \ll 1$). This requirement is, for example, frequently satisfied for antennas.

The exponential function $e^{-ik\boldsymbol{n}\cdot\boldsymbol{x}}$ can be expanded in a power series. This corresponds to a *decomposition of the radiation into multipoles* (dipole radiation, quadrupole radiation, etc.). In the first approximation, one sets

$$e^{-ik(\boldsymbol{n}\cdot\boldsymbol{x})} = 1\,.$$

(That is, the retardation within the current system is neglected.) It then follows that

$$\begin{aligned} \boldsymbol{E} &= ik\frac{e^{ikR}}{R}\frac{1}{c}\int \boldsymbol{i}_\perp\,\mathrm{d}V\,, \\ \boldsymbol{H} &= ik\frac{e^{ikR}}{R}\frac{1}{c}\boldsymbol{n}\times\int \boldsymbol{i}\,\mathrm{d}V\,. \end{aligned} \qquad [28.19]$$

This is called *dipole radiation.* In this case, the dependence of the field intensities on **n** can be given explicitly. Let

$$\boldsymbol{j} = \int \boldsymbol{i}\,\mathrm{d}V\,. \qquad [28.20]$$

Then,

$$\boldsymbol{E} = \frac{ik}{c}\frac{e^{ikR}}{R}\boldsymbol{j}_\perp, \qquad \boldsymbol{H} = \boldsymbol{n}\times\boldsymbol{E}. \tag{28.21}$$

For the case of a *linear oscillation* parallel to the z axis,

$$|j_\perp| = |j|\sin\vartheta .$$

The Poynting vector $(\boldsymbol{S} = (c/4\pi)|\boldsymbol{E}|^2\boldsymbol{n})$ is then

$$S_n = \frac{c}{4\pi}\frac{k^2}{c^2R^2}j_0^2\sin^2\vartheta\sin^2(kR-\omega t+\alpha)$$

$$= \frac{1}{4\pi R^2}\frac{1}{c^3}\left(\frac{\mathrm{d}\boldsymbol{j}}{\mathrm{d}t}\right)^2_{\mathrm{ret}}\sin^2\vartheta .$$

The energy radiated per second is obtained by integrating over the surface of a sphere:

$$\int S_n\,\mathrm{d}f = \frac{2}{3}\frac{1}{c^3}\left(\frac{\mathrm{d}\boldsymbol{j}}{\mathrm{d}t}\right)^2_{\mathrm{ret}}, \tag{28.22}$$

since

$$\mathrm{d}f = R^2\,\mathrm{d}\Omega, \qquad \frac{1}{4\pi}\int\sin^2\vartheta\,\mathrm{d}\Omega = \frac{2}{3}, \qquad \frac{1}{4\pi}\int\cos^2\vartheta\,\mathrm{d}\Omega = \frac{1}{3}.$$

The case of a radiating system consisting of a *point charge* is contained in the above if the charge oscillates as a unit:

$$\boldsymbol{i} = \varrho\boldsymbol{v}, \qquad \boldsymbol{j} = \int\boldsymbol{i}\,\mathrm{d}V = e\boldsymbol{v}.$$

Thus, for an oscillating point charge,

$$\int S_n\,\mathrm{d}f = -\frac{\mathrm{d}E}{\mathrm{d}t} = \frac{2}{3}\frac{e^2}{c^3}\dot{\boldsymbol{v}}^2_{\mathrm{ret}}, \tag{28.23}$$

if λ is large as compared with the linear dimensions of the region in which the charge moves. Naturally, λ must also be large as compared with the dimensions of the charge: $\lambda \gg a$. For an electron, $a \sim e^2/mc^2 \sim 10^{-13}$ cm.

It is often useful to introduce a polarization vector $\boldsymbol{P}$:

$$\boldsymbol{i}=\frac{\partial \boldsymbol{P}}{\partial t}, \qquad \varrho=-\operatorname{div} \boldsymbol{P}. \qquad [28.24]$$

For periodic processes, then, $\boldsymbol{P}=i\,\boldsymbol{i}/\omega$ and

$$\boldsymbol{E}=-\frac{e^{ikR}}{R}\frac{1}{c^2}\int\frac{\partial^2 \boldsymbol{P}_\perp}{\partial t^2}\,\mathrm{d}V=-\frac{e^{ikR}}{R}\frac{1}{c^2}\frac{\mathrm{d}^2\boldsymbol{\Pi}_\perp}{\mathrm{d}t^2}, \qquad [28.25]$$

where

$$\boldsymbol{\Pi}=\int\boldsymbol{P}\,\mathrm{d}V, \qquad \boldsymbol{j}=\frac{\mathrm{d}\boldsymbol{\Pi}}{\mathrm{d}t}. \qquad [28.26]$$

Nonlinear oscillations. Let $\boldsymbol{\Pi}_1$ and $\boldsymbol{\Pi}_2$ be two orthogonal linear components of the oscillation. Then,

$$\boldsymbol{E}_1=-\frac{e^{ikR}}{R}\frac{1}{c^2}\ddot{\boldsymbol{\Pi}}_{1\perp}=\boldsymbol{H}_1\times\boldsymbol{n},$$

$$\boldsymbol{E}_2=-\frac{e^{ikR}}{R}\frac{1}{c^2}\ddot{\boldsymbol{\Pi}}_{2\perp}=\boldsymbol{H}_2\times\boldsymbol{n}.$$

The Poynting vector is then not additive:

$$\boldsymbol{S}=\boldsymbol{S}_1+\boldsymbol{S}_2+\frac{c}{4\pi}\{\boldsymbol{E}_1\times\boldsymbol{H}_2+\boldsymbol{E}_2\times\boldsymbol{H}_1\}.$$

However, the cross terms vanish when integrated over all directions. That is, their component in the $\boldsymbol{n}$ direction is

$$\frac{c}{4\pi}(\boldsymbol{n}\cdot\boldsymbol{E}_1\times\boldsymbol{H}_2+\boldsymbol{n}\cdot\boldsymbol{E}_2\times\boldsymbol{H}_1)$$

$$=\frac{c}{4\pi}(\boldsymbol{E}_1\cdot\boldsymbol{H}_2\times\boldsymbol{n}+\boldsymbol{E}_2\cdot\boldsymbol{H}_1\times\boldsymbol{n})$$

$$=\frac{c}{4\pi}2(\boldsymbol{E}_1\cdot\boldsymbol{E}_2)=\text{constant}\,[\boldsymbol{\Pi}_1-\boldsymbol{n}(\boldsymbol{n}\cdot\boldsymbol{\Pi}_1)]\cdot[\boldsymbol{\Pi}_2-\boldsymbol{n}(\boldsymbol{n}\cdot\boldsymbol{\Pi}_2)]$$

$$=\text{constant}\,[(\boldsymbol{\Pi}_1\cdot\boldsymbol{\Pi}_2)-(\boldsymbol{n}\cdot\boldsymbol{\Pi}_1)(\boldsymbol{n}\cdot\boldsymbol{\Pi}_2)],$$

where constant implies independent of $\boldsymbol{n}$. We now use the provision that $\boldsymbol{\Pi}_1$ is perpendicular to $\boldsymbol{\Pi}_2$. With $\boldsymbol{\Pi}_1$ parallel to the x direction and $\boldsymbol{\Pi}_2$ parallel to the y direction, our expression becomes

$$\text{constant} \times n_x n_y \, .$$

This vanishes when integrated over the surface of a sphere. We thus have the important relation

$$\oint S_n \, \mathrm{d}f = \oint S_{1n} \, \mathrm{d}f + \oint S_{2n} \, \mathrm{d}f \, , \qquad [28.27]$$

if the oscillations are mutually perpendicular.

Special case: circular oscillations. This case is important in optics. Here, linearly polarized light is emitted along the x and y directions, and circularly polarized light along the z

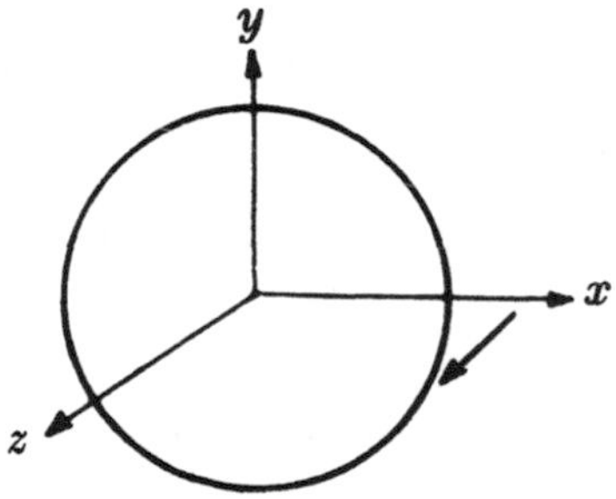

Figure 28.2

direction. In any other abitrary direction, the light is elliptically polarized.

3. *The Hertz vector.* In the radiation zone, we had the condition $kR \gg 1$ (or $R \gg \lambda$). We will no longer make this restriction. Instead, we now assume that

$$d \ll \lambda \, , \qquad d \ll R \, ,$$

for arbitrary kR. This type of radiation is also called dipole radiation, although now we will no longer limit our considerations to the radiation zone.

In Eqs. [28.12] and [28.13] we can once again replace r_{PQ}

by an average value R. Then,

$$\boldsymbol{A} = \frac{1}{c}\frac{e^{ikR}}{R}\int \boldsymbol{i}\,\mathrm{d}V = \frac{1}{c}\frac{e^{ikR}}{R}\boldsymbol{j}.$$

In order to show the connection with the usual notation, we set

$$\boldsymbol{j} = \frac{\mathrm{d}\boldsymbol{\Pi}}{\mathrm{d}t} = -i\omega\boldsymbol{\Pi},$$

so that

$$\frac{1}{c}\boldsymbol{j} = -ik\boldsymbol{\Pi}$$

and

$$\boldsymbol{A} = -ik\frac{e^{ikR}}{R}\boldsymbol{\Pi}. \qquad [28.28]$$

Because of [28.3′], we obtain

$$\varphi = -\operatorname{div}\left(\frac{e^{ikR}}{R}\boldsymbol{\Pi}\right) = -\boldsymbol{\Pi}\cdot\operatorname{grad}\frac{e^{ikR}}{R}, \qquad [28.29]$$

since $\boldsymbol{\Pi}$ is independent of position.

The field intensities are now calculated without neglecting powers of $1/kR$. It can already be seen that this solution is related to a dipole, since for $k=0$, φ becomes the electrostatic potential for a dipole of moment $\boldsymbol{\Pi}$. If, however, the dipole oscillates, additional terms also contribute significantly. We have

$$\boldsymbol{H} = \operatorname{curl}\boldsymbol{A} = -ik\operatorname{grad}\frac{e^{ikR}}{R}\times\boldsymbol{\Pi}.$$

Because

$$\operatorname{grad} f(R) = \frac{\mathrm{d}f}{\mathrm{d}R}\cdot\boldsymbol{n},$$

where $\boldsymbol{n} = \boldsymbol{x}/R$, then

$$\boldsymbol{H} = -ik\frac{\mathrm{d}}{\mathrm{d}R}\left(\frac{e^{ikR}}{R}\right)(\boldsymbol{n}\times\boldsymbol{\Pi}) = -ik\frac{e^{ikR}}{R}\left(ik - \frac{1}{R}\right)\boldsymbol{n}\times\boldsymbol{\Pi}.$$

Thus,

$$\boldsymbol{H}=\frac{e^{ikR}}{R}\left(k^2+\frac{ik}{R}\right)\boldsymbol{n}\times\boldsymbol{\Pi}. \qquad [28.30]$$

The first term has already been obtained in the radiation zone. The second term is large at small distances. Then φ becomes

$$\varphi=-\frac{\mathrm{d}}{\mathrm{d}R}\left(\frac{e^{ikR}}{R}\right)(\boldsymbol{n}\cdot\boldsymbol{\Pi})=\left(-\frac{ik}{R^2}+\frac{1}{R^3}\right)e^{ikR}\boldsymbol{x}\cdot\boldsymbol{\Pi}.$$

Calculating $\operatorname{grad}\varphi$ yields

$$\operatorname{grad}\varphi=\frac{e^{ikR}}{R}\left\{\left(-\frac{ik}{R}+\frac{1}{R^2}\right)\boldsymbol{\Pi}+\left(k^2+\frac{3ik}{R}-\frac{3}{R^2}\right)\boldsymbol{n}(\boldsymbol{n}\cdot\boldsymbol{\Pi})\right\}.$$

Thus,

$$\boldsymbol{E}=ik\cdot\boldsymbol{A}-\operatorname{grad}\varphi=\frac{e^{ikR}}{R}k^2\boldsymbol{\Pi}-\operatorname{grad}\varphi$$

$$=\frac{e^{ikR}}{R}\left\{\left(k^2+\frac{ik}{R}-\frac{1}{R^2}\right)\boldsymbol{\Pi}+\left(-k^2-\frac{3ik}{R}+\frac{3}{R^2}\right)\boldsymbol{n}(\boldsymbol{n}\cdot\boldsymbol{\Pi})\right\}. \qquad [28.31]$$

This solution for $\boldsymbol{H}$ and $\boldsymbol{E}$ is known under the name of the Hertz *oscillating dipole.*

The possibility of passing to the limit $\boldsymbol{d}\rightarrow 0$ exists. Our equations are then solutions of the homogeneous equations ($\varrho=0$, $\boldsymbol{i}=0$) everywhere except at the origin, at which point a singularity exists.

For the case where $R\ll\lambda\,(kR\ll 1)$, one has

$$\boldsymbol{E}=\frac{1}{R^3}\{3\boldsymbol{n}(\boldsymbol{n}\cdot\boldsymbol{\Pi})-\boldsymbol{\Pi}\}.$$

This is the field of an *electrostatic dipole* of moment $\boldsymbol{\Pi}$. (This can be immediately seen by considering the z axis to lie along the $\boldsymbol{n}$ direction.)

For $\boldsymbol{H}$ there results

$$\boldsymbol{H}=\frac{ik}{R^2}\,\boldsymbol{n}\times\boldsymbol{\Pi}=-\frac{1}{cR^2}\,\boldsymbol{n}\times\boldsymbol{j}=+\frac{1}{c}\frac{\boldsymbol{j}\times\boldsymbol{n}}{R^2}\,.$$

This is simply the Biot-Savart field corresponding to the current $\boldsymbol{j}$.

In the radiation zone where $R\gg\lambda\,(kR\gg 1)$, we obtain the previous results:

$$\boldsymbol{H}=k^2\,\frac{e^{ikR}}{R}\,\boldsymbol{n}\times\boldsymbol{\Pi}\,,$$

$$\boldsymbol{E}=k^2\,\frac{e^{ikR}}{R}\,\{\boldsymbol{\Pi}-\boldsymbol{n}(\boldsymbol{n}\cdot\boldsymbol{\Pi})\}\,.$$

d. Integration of the inhomogeneous wave equations. Second method

Almost all of optics can be treated with the method of periodic time dependence. However, we will demonstrate still another method for integrating the inhomogeneous wave equation

$$\nabla^2\varphi-\frac{1}{c^2}\frac{\partial^2\varphi}{\partial t^2}=-4\pi\varrho(\boldsymbol{x},t)\,. \qquad [28.32]$$

In this second method, no decomposition into Fourier integrals or Fourier series in time is made.

A *new variable* τ is introduced:

$$t=\tau-\frac{r}{c}, \qquad [28.33]$$

where $r=|\boldsymbol{x}_P-\boldsymbol{x}_Q|$. We also employ polar coordinates r, ϑ, ψ. Thus,

$$\varrho(\boldsymbol{x};t)=\varrho\left(r,\vartheta,\psi;\tau-\frac{r}{c}\right).$$

We denote

$$\frac{\partial F}{\partial r}\equiv\left(\frac{\partial F}{\partial r}\right)_t \quad \text{(partial differentiation at constant } t\text{)}\,,$$
$$\frac{\mathrm{d}F}{\mathrm{d}r}\equiv\left(\frac{\partial F}{\partial r}\right)_\tau \quad \text{(partial differentiation at constant } \tau\text{)}\,. \qquad [28.34]$$

Then,

$$\frac{\partial F}{\partial r} = \frac{\mathrm{d}F}{\mathrm{d}r} + \frac{1}{c}\frac{\partial F}{\partial t}, \qquad \frac{\partial}{\partial t} = \frac{\partial}{\partial \tau}.$$

The divergence of a vector $\boldsymbol{A}$ becomes

$$\mathrm{div}_\tau \boldsymbol{A} = \boldsymbol{A}\frac{1}{r^2}\frac{\mathrm{d}}{\mathrm{d}r}(r^2 A_r) + (\text{derivatives with respect to } \vartheta \text{ and } \psi) \quad \text{at fixed } \tau,$$

$$\mathrm{div}_t \boldsymbol{A} = \frac{1}{r^2}\frac{\partial}{\partial r}(r^2 A_r) + (\text{derivatives with respect to } \vartheta \text{ and } \psi) \quad \text{at fixed } t.$$

According to the above relations,

$$\mathrm{div}_\tau \boldsymbol{A} = \mathrm{div}_t \boldsymbol{A} - \frac{1}{c}\frac{\partial A_r}{\partial t}.$$

Application to $\boldsymbol{A} = (1/r)\,\boldsymbol{\nabla}_t \varphi$, where $\boldsymbol{\nabla}_t$ represents the gradient at fixed t, yields

$$\begin{aligned}\mathrm{div}_\tau\left(\frac{1}{r}\boldsymbol{\nabla}_t\varphi\right) &= \mathrm{div}_t\left(\frac{1}{r}\boldsymbol{\nabla}_t\varphi\right) - \frac{1}{rc}\frac{\partial^2\varphi}{\partial t\,\partial r},\\ &= \frac{1}{r}\nabla_t^2\varphi + \left(\boldsymbol{\nabla}_t\frac{1}{r}\right)(\boldsymbol{\nabla}_t\varphi) - \frac{1}{rc}\frac{\partial^2\varphi}{\partial t\,\partial r},\\ &= \frac{1}{r}\nabla_t^2\varphi - \frac{1}{r^2}\frac{\partial\varphi}{\partial r} - \frac{1}{rc}\frac{\partial^2\varphi}{\partial t\,\partial r}.\end{aligned}$$

During integration, we shall require derivatives at fixed τ; thus we replace $\partial/\partial r$ by $\mathrm{d}/\mathrm{d}r$. Then,

$$\begin{aligned}\mathrm{div}_\tau\left(\frac{1}{r}\boldsymbol{\nabla}_t\varphi\right) &= \frac{1}{r}\nabla_t^2\varphi - \frac{1}{r^2}\frac{\mathrm{d}\varphi}{\mathrm{d}r} - \frac{1}{r^2}\frac{1}{c}\frac{\partial\varphi}{\partial t} - \frac{1}{rc}\frac{\partial}{\partial t}\frac{\mathrm{d}\varphi}{\mathrm{d}r} - \frac{1}{rc^2}\frac{\partial^2\varphi}{\partial t^2}\\ &= \frac{1}{r}\left\{\nabla_t^2\varphi - \frac{1}{c^2}\frac{\partial^2\varphi}{\partial t^2}\right\} - \frac{1}{r^2}\frac{\mathrm{d}}{\mathrm{d}r}\left\{\varphi + r\frac{1}{c}\frac{\partial\varphi}{\partial t}\right\}. \qquad [28.35]\end{aligned}$$

We will integrate this equation (at *fixed* τ) over a volume from which the point P at $r=0$ is excluded by means of

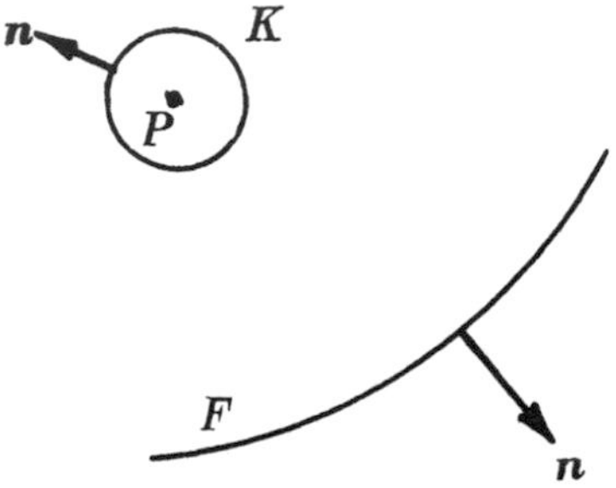

Figure 28.3

a small sphere K (Fig. 28.3). Thus,

$$\int_{F+K} \frac{1}{r}\left(\frac{\partial \varphi}{\partial n}\right)_t \mathrm{d}f = -4\pi \int \frac{\varrho_Q(r, \vartheta, \psi; \tau - r/c)}{r} \mathrm{d}V_Q$$

$$-\int_F \left(\varphi + r\frac{1}{c}\frac{\partial \varphi}{\partial t}\right) \mathrm{d}\Omega + \int_K \left(\varphi + r\frac{1}{c}\frac{\partial \varphi}{\partial t}\right) \mathrm{d}\Omega,$$

where $\mathrm{d}\Omega$ is the element of solid angle and $\mathrm{d}V_Q = r^2 \mathrm{d}r \mathrm{d}\Omega$. In the limit $K \to 0$, the integral over the small sphere on the left-hand side vanishes, while that on the right-hand side gives $4\pi\varphi_P$. With the aid of the geometrical relation $\mathrm{d}\Omega = (1/r^2)\cos(\boldsymbol{n}\cdot\boldsymbol{r})\mathrm{d}f$, which is valid for any arbitrary surface F, there results the important formula

$$\varphi_P(\tau) = \int \frac{\varrho_Q(r, \vartheta, \psi; \tau - r/c)}{r} \mathrm{d}V_Q$$

$$+ \frac{1}{4\pi}\int_F \mathrm{d}f \left\{\frac{1}{r}\left(\frac{\partial \varphi}{\partial n}\right)_t + \left(\frac{\varphi}{r^2} + \frac{1}{rc}\frac{\partial \varphi}{\partial t}\right)\cos(\boldsymbol{n}, \boldsymbol{r})\right\}. \qquad [28.36]$$

In this relation, τ is fixed and $r \equiv r_{PQ}$.

This equation is also valid as a solution of the homogeneous wave equation

$$\nabla^2\varphi - \frac{1}{c^2}\frac{\partial^2\varphi}{\partial t^2} = 0 .$$

In this case,

$$4\pi\varphi_P = \int_F \mathrm{d}f \left\{ \frac{1}{r}\left(\frac{\partial\varphi}{\partial n}\right)_t + \left(\frac{\varphi}{r^2} + \frac{1}{rc}\frac{\partial\varphi}{\partial t}\right)\cos(\boldsymbol{n}, \boldsymbol{r}) \right\} . \qquad [28.37]$$

If the surface F is a sphere, then $\cos(\boldsymbol{n}, \boldsymbol{r}) = 1$, $\mathrm{d}f = r^2\mathrm{d}\Omega$, $(\partial/\partial \boldsymbol{n})_t = \partial/\partial \boldsymbol{r}$, and there follows

$$4\pi\varphi_P = \int_F \mathrm{d}\Omega \left\{ \varphi + r\left(\frac{\partial\varphi}{\partial r} + \frac{1}{c}\frac{\partial\varphi}{\partial t}\right) \right\} . \qquad [28.38]$$

For this case,

$$\lim_{r\to\infty} \varphi_P = 0 .$$

If, in addition, the *radiation condition*

$$\lim_{r\to\infty} r\left(\frac{\partial\varphi}{\partial r} + \frac{1}{c}\frac{\partial\varphi}{\partial t}\right) = 0 \qquad [28.39]$$

is satisfied, then the surface F can be taken at infinity so that the surface integral vanishes.

For $\varphi = \varphi_0 e^{-i\omega t}$, $k = \omega/c$, the previous relations are once again obtained.

29. THE FIELD OF A POINT CHARGE IN UNIFORM MOTION

The consideration of this case provides preparation for the special theory of relativity.

We start with the equations

$$\operatorname{div} \boldsymbol{A} + \frac{1}{c}\dot{\varphi} = 0 ,$$

$$\nabla^2\varphi - \frac{1}{c^2}\ddot{\varphi} = -4\pi\varrho ,$$

$$\nabla^2\boldsymbol{A} - \frac{1}{c^2}\ddot{\boldsymbol{A}} = -\frac{4\pi}{c}\boldsymbol{i} .$$

Let the charge ϱ move with velocity v in the x direction so that $x_{\varrho} = vt$. We seek a solution of the Maxwell equations which depends only upon the position of the field point relative to the position of the charge, $(x - vt, y, z)$. The time dependence, therefore, should appear only in the form $x - vt$. Then,

$$\frac{\partial}{\partial t} = -v \frac{\partial}{\partial x} .$$

From

$$i_x = v\varrho, \qquad i_y = i_z = 0 ,$$

it follows that

$$A_y = A_z = 0 , \qquad \frac{\partial A_x}{\partial x} - \frac{v}{c}\frac{\partial \varphi}{\partial x} = 0 .$$

This is satisfed by the expression

$$A_x = (v/c)\varphi .$$

Only the equation for φ,

$$\left(1 - \frac{v^2}{c^2}\right)\frac{\partial^2 \varphi}{\partial x^2} + \frac{\partial^2 \varphi}{\partial y^2} + \frac{\partial^2 \varphi}{\partial z^2} = -4\pi\varrho , \qquad [29.1]$$

must still be satisfied. The factor $(1 - v^2/c^2)$ implies a contraction. In order to obtain the same result as for the static case, we introduce

$$x' = \frac{x - vt}{\sqrt{1 - v^2/c^2}} . \qquad [29.2]$$

The equation for φ then becomes

$$\frac{\partial^2 \varphi}{\partial x'^2} + \frac{\partial^2 \varphi}{\partial y^2} + \frac{\partial^2 \varphi}{\partial z^2} = -4\pi\varrho , \qquad [29.3]$$

which has the solution

$$\varphi = \frac{e}{r'} . \qquad [29.4]$$

Here, however,

$$r'^2 = \frac{(x - vt)^2}{1 - v^2/c^2} + y^2 + z^2 . \qquad [29.5]$$

For the electric field it follows that

$$\boldsymbol{E} = -\frac{1}{c}\dot{\boldsymbol{A}} - \operatorname{grad}\varphi ;$$

$$E_x = \frac{v^2}{c^2}\frac{\partial\varphi}{\partial x} - \frac{\partial\varphi}{\partial x} = -\left(1 - \frac{v^2}{c^2}\right)\frac{\partial\varphi}{\partial x},$$

$$E_y = -\frac{\partial\varphi}{\partial y}, \qquad E_z = -\frac{\partial\varphi}{\partial z};$$

$$\left.\begin{aligned} E_x &= \frac{e}{r'^2}\frac{x - vt}{r'} \\ E_y &= \frac{e}{r'^2}\frac{y}{r'} \\ E_z &= \frac{e}{r'^2}\frac{z}{r'} \end{aligned}\right\} \quad \boldsymbol{E} = \frac{e}{r'^3}\boldsymbol{r} . \qquad [29.6]$$

Therefore $\boldsymbol{E}$ is in the direction of $\boldsymbol{r} = (x - vt, y, z)$ and not of $\boldsymbol{r}' = (x', y, z)$. The equipotential surfaces $r' =$ constant are flattened ellipsoids of rotation (***Heaviside ellipsoids***). The vector $\boldsymbol{E}$ is not normal to these equipotential surfaces.

For the magnetic field it follows that

$$H_x = 0, \qquad H_y = +\frac{\partial A_x}{\partial z} = +\frac{v}{c}\frac{\partial\varphi}{\partial z} = -\frac{v}{c}E_z,$$

$$H_z = -\frac{\partial A_x}{\partial y} = -\frac{v}{c}\frac{\partial\varphi}{\partial y} = +\frac{v}{c}E_y .$$

With $\boldsymbol{v} = (v, 0, 0)$, this can be written as

$$\boldsymbol{H} = \frac{1}{c}\boldsymbol{v}\times\boldsymbol{E} \qquad [29.7]$$

and

$$\boldsymbol{A} = \frac{e\boldsymbol{v}}{c \cdot r'} . \qquad [29.8]$$

These relations are the precursors of a much more general result; The Maxwell equations are invariant under a group of transformations (the *Lorentz group*) which transform between coordinate systems in uniform motion and have the property that

$$x'^2 + y'^2 + z'^2 - c^2 \cdot t'^2 = x^2 + y^2 + z^2 - c^2 \cdot t^2 .$$

For a special choice of coordinates, such a transformation has the form

$$x' = \frac{x - vt}{\sqrt{1 - v^2/c^2}}, \qquad y' = y , \qquad z' = z , \qquad t' = \frac{t - (v/c^2)x}{\sqrt{1 - v^2/c^2}} .$$

The field intensities also transform in such a manner that the Maxwell equations remain unchanged. The laws of nature are thus the same in all reference systems. What is new is the fact that the time also undergoes a transformation. In Newtonian and Galilean physics one has

$$x' = x - vt , \qquad y' = y , \qquad z' = z , \qquad t' = t .$$

Here, on the contrary, two events at different space points which are simultaneous in one reference system are not simultaneous in another. This is very closely related to the finite propagation velocity of electromagnetic fields.

30. RADIATION DAMPING

We have seen (Eq. [28.23]) that the energy radiated per second by an oscillating point charge is given by

$$-\frac{\mathrm{d}E}{\mathrm{d}t} = \frac{2}{3}\frac{e^2}{c^3}\dot{\boldsymbol{v}}^2_{\mathrm{ret}} . \qquad [30.1]$$

According to the law of conservation of energy, the field must exert a reactive force on the charge. This force can be guessed at by considering the fact that

$$\frac{\mathrm{d}E}{\mathrm{d}t} = \boldsymbol{K} \cdot \boldsymbol{v} .$$

Since, however, the energy fluctuates, one need only demand that this be true for the time-averaged quantities:

$$\overline{\boldsymbol{K}\cdot\boldsymbol{v}} = \overline{\frac{\mathrm{d}E}{\mathrm{d}t}} = -\frac{2}{3}\frac{e^2}{c^3}\overline{\dot{\boldsymbol{v}}^2} = -\frac{2}{3}\frac{e^2}{c^3}\overline{\frac{\mathrm{d}}{\mathrm{d}t}(\dot{\boldsymbol{v}}\cdot\boldsymbol{v})} + \frac{2}{3}\frac{e^2}{c^3}\overline{\ddot{\boldsymbol{v}}\cdot\boldsymbol{v}}\,.$$

The first term on the right vanishes, since it is a total derivative. Thus, the *radiation reaction* (the *radiation damping force*) is obtained:

$$\boldsymbol{K} = +\frac{2}{3}\frac{e^2}{c^3}\ddot{\boldsymbol{v}} = \frac{2}{3}\frac{e^2}{c^3}\dddot{\boldsymbol{x}}\,. \qquad [30.2]$$

This is only a heuristic consideration. However, this force can be derived directly from the reaction of the field on an electron. For this purpose, the charge must be assumed to have a finite extent. It is found that Eq. [30.2] is correct if terms of the order of $(v/c)^2$ are neglected.

Calculation of the self-force on a charge of finite extent [A-4]

The electron will be represented by a charge distribution in motion, $(\varrho, \boldsymbol{i})$, regarding which certain special assumptions will be made.

The field produced at an arbitrary point P by the electron is given by

$$\boldsymbol{E}_P = -\frac{1}{c}\dot{\boldsymbol{A}}_P - \operatorname{grad}_P \varphi_P\,, \qquad [30.3]$$

$$\boldsymbol{H}_P = \operatorname{curl}_P \boldsymbol{A}_P\,, \qquad [30.4]$$

where

$$\varphi_P(t) = \int \varrho_Q\left(t - \frac{r_{PQ}}{c}\right)\frac{\mathrm{d}V_Q}{r_{PQ}}\,, \qquad [30.5]$$

$$\boldsymbol{A}_P(t) = \frac{1}{c}\int \boldsymbol{i}_Q\left(t - \frac{r_{PQ}}{c}\right)\frac{\mathrm{d}V_Q}{r_{PQ}}\,. \qquad [30.6]$$

We are interested in the field existing in the region of the particle itself. Since this region is small, the retardation will not play a major role. We expand in powers of $-r/c$:

$$\varphi_P = \int \frac{1}{r_{PQ}} \sum_{n=0}^{\infty} \frac{1}{n!} \left(-\frac{r_{PQ}}{c}\right)^n \frac{\partial^n \varrho_Q}{\partial t^n} \mathrm{d}V_Q$$

$$= \sum_{n=0}^{\infty} \frac{1}{n!} \left(-\frac{1}{c}\right)^n \frac{\partial^n}{\partial t^n} \int r_{PQ}^{n-1} \varrho_Q \,\mathrm{d}V_Q \,,$$

$$\boldsymbol{A}_P = \frac{1}{c} \sum_{n=0}^{\infty} \frac{1}{n!} \left(-\frac{1}{c}\right)^n \frac{\partial^n}{\partial t^n} \int r_{PQ}^{n-1} \boldsymbol{i}_Q \,\mathrm{d}V_Q \,.$$

In the expression for φ, the $n=0$ term is simply the Coulomb potential. The $n=1$ term vanishes since $\int \varrho_Q \,\mathrm{d}V_Q = e =$ constant and $\partial(\text{constant})/\partial t = 0$. In the remainder of the series we replace n by $n+2$. Thus,

$$\varphi_P = \int \frac{\varrho_Q}{r_{PQ}} \mathrm{d}V_Q + \sum_{n=0}^{\infty} \frac{1}{(n+2)!} \left(-\frac{1}{c}\right)^{n+2} \frac{\partial^{n+2}}{\partial t^{n+2}} \int r_{PQ}^{n+1} \varrho_Q \,\mathrm{d}V_Q \,.$$

Here, $\partial/\partial t$ operates only on ϱ_Q. Using the continuity equation

$$\frac{\partial \varrho_Q}{\partial t} = -\operatorname{div}_Q \boldsymbol{i}_Q \,,$$

the identity of Eq. [4.10], Gauss's law, and the fact that under the integral sign,

$$\boldsymbol{i}_Q \cdot \operatorname{grad}_Q = -\boldsymbol{i}_Q \cdot \operatorname{grad}_P \,,$$

we get

$$\varphi_P = \int \frac{\varrho_Q}{r_{PQ}} \mathrm{d}V_Q - \frac{1}{c^2} \sum_{n=0}^{\infty} \frac{1}{(n+2)!} \left(-\frac{1}{c}\right)^n \frac{\partial^{n+1}}{\partial t^{n+1}} \int \boldsymbol{i}_Q \cdot \operatorname{grad}_P r_{PQ}^{n+1} \,\mathrm{d}V_Q \,.$$

The electric field is then found from Eq. [28.2]:

$$\boldsymbol{E}_P = -\operatorname{grad}_P \int \frac{\varrho_Q}{r_{PQ}} \mathrm{d}V_Q - \frac{1}{c^2} \sum_{n=0}^{\infty} \left(-\frac{1}{c}\right)^n \frac{\partial^{n+1}}{\partial t^{n+1}}$$

$$\times \int \left\{ \frac{1}{n!} \boldsymbol{i}_Q r_{PQ}^{n-1} - \frac{1}{(n+2)!} \operatorname{grad}_P (\boldsymbol{i}_Q \cdot \operatorname{grad}_P r_{PQ}^{n+1}) \right\} \mathrm{d}V_Q \,. \qquad [30.7]$$

In order to calculate the force exerted by the field on the particle we make the following assumptions:

1. The charge distribution is rigid and moves only with

translational motion (although not uniformly). That is,

$$\boldsymbol{i}(\boldsymbol{x}, t) = \varrho(\boldsymbol{x}, t)\,\boldsymbol{v}(t)\,, \tag{30.8}$$

where $\boldsymbol{v}(t)$ is independent of $\boldsymbol{x}$.

2. The reference system is chosen so that $v/c \ll 1$. We keep only terms of zero order in v/c. This places no restrictive assumptions on the magnitude of the acceleration or on higher time derivatives of $\boldsymbol{v}$. Indeed, our derivation is exact in the instantaneous rest system of the electron. Furthermore, terms of first order in v/c vanish so that the first neglected terms are of order $(v/c)^2$.

According to Eqs. [26.1] and [30.8], the force per unit volume is

$$\boldsymbol{k} = \varrho\left(\boldsymbol{E} + \frac{1}{c}\,\boldsymbol{v}\times\boldsymbol{H}\right).$$

As the total force on the electron there remains

$$\boldsymbol{K} = \int \varrho_P \boldsymbol{E}_P \,\mathrm{d}V_P\,,$$

since the magnetic Lorentz force is of order $(v/c)^2$. If the expansion of Eq. [30.7] is substituted into the above expression for $\boldsymbol{K}$, it is seen that the Coulomb term drops out since

$$\iint \varrho_P \varrho_Q \frac{\boldsymbol{x}_P - \boldsymbol{x}_Q}{r_{PQ}^3}\,\mathrm{d}V_P\,\mathrm{d}V_Q = 0$$

(as can be seen by interchanging the variables of integration). There remains

$$\boldsymbol{K} = \sum_{n=0}^{\infty} \boldsymbol{K}^{(n)}\,, \tag{30.9}$$

where

$$\boldsymbol{K}^{(n)} = -\left(-\frac{1}{c}\right)^{n+2} \int \varrho_P \,\mathrm{d}V_P \frac{\partial^{n+1}}{\partial t^{n+1}}$$
$$\times \int \varrho_Q \left\{\frac{1}{n!}\,\boldsymbol{v} r_{PQ}^{n-1} - \frac{1}{(n+2)!}\,\mathrm{grad}_P\,(\boldsymbol{v}\cdot \mathrm{grad}_P r_{PQ}^{n+1})\right\} \mathrm{d}V_Q\,. \tag{30.10}$$

The principal terms are $\boldsymbol{K}^{(0)}$ and $\boldsymbol{K}^{(1)}$.

In the calculation of $K^{(0)}$, a term with $\partial\varrho_Q/\partial t$ appears. Because of Eqs. [24.5] and [30.8], this is equal to $-\mathrm{div}_Q(\varrho_Q \boldsymbol{v})$. This term,

$$+\frac{1}{c^2}\iint \varrho_P \,\mathrm{div}_Q(\varrho_Q \boldsymbol{v})\left\{\boldsymbol{v}\frac{1}{r_{PQ}}-\frac{1}{2}\,\mathrm{grad}_P(\boldsymbol{v}\cdot\mathrm{grad}_P r_{PQ})\right\}\mathrm{d}V_P\,\mathrm{d}V_Q\,,$$

can be transformed into

$$-\frac{1}{c^2}\iint \varrho_P\,\varrho_Q\,\boldsymbol{v}\cdot\mathrm{grad}_Q\left\{\boldsymbol{v}\frac{1}{r_{PQ}}-\frac{1}{2}\,\mathrm{grad}_P(\boldsymbol{v}\cdot\mathrm{grad}_P r_{PQ})\right\}\mathrm{d}V_P\,\mathrm{d}V_Q$$

by using the integral theorem of Eq. [8.7]. Since the differential operators affect only r_{PQ}, grad_Q can be replaced by $-\,\mathrm{grad}_P$ and vice versa, and there results

$$+\frac{1}{c^2}\iint \varrho_P\varrho_Q\,\boldsymbol{v}\cdot\mathrm{grad}_P\left\{\boldsymbol{v}\frac{1}{r_{PQ}}-\frac{1}{2}\,\mathrm{grad}_Q(\boldsymbol{v}\cdot\mathrm{grad}_Q r_{PQ})\right\}\mathrm{d}V_P\,\mathrm{d}V_Q\,.$$

The same result—although with the opposite sign—would be obtained by interchanging the variables of integration. This shows that the term must be zero. There remains for $\boldsymbol{K}^{(0)}$,

$$\boldsymbol{K}^{(0)}=-\frac{1}{c^2}\iint \varrho_P\,\varrho_Q\left\{\dot{\boldsymbol{v}}\frac{1}{r_{PQ}}-\frac{1}{2}\,\mathrm{grad}_P(\dot{\boldsymbol{v}}\cdot\mathrm{grad}_P r_{PQ})\right\}\mathrm{d}V_P\,\mathrm{d}V_Q\,.$$

This integral can be simplified by assuming a *spherically symmetric charge distribution.* The integral of

$$\frac{\partial^2 r_{PQ}}{\partial x_{P_i}\partial x_{P_k}}=\delta_{ik}\frac{1}{r_{PQ}}-\frac{(x_{P_i}-x_{Q_i})(x_{P_k}-x_{Q_k})}{r_{PQ}^3}$$

vanishes for $i\neq k$, and $\partial^2 r_{PQ}/\partial x_{P_i}^2$ can be replaced by $\nabla^2 r/3=2/3r$. Hence,

$$\mathrm{grad}_P(\dot{\boldsymbol{v}}\cdot\mathrm{grad}_P r_{PQ})\to\frac{2}{3}\dot{\boldsymbol{v}}\frac{1}{r_{PQ}}$$

and

$$\boldsymbol{K}^{(0)}=-\,\mu\dot{\boldsymbol{v}}\,, \qquad [30.11]$$

where

$$\mu=\frac{2}{3}\frac{1}{c^2}\iint\frac{\varrho_P\,\varrho_Q}{r_{PQ}}\,\mathrm{d}V_P\,\mathrm{d}V_Q\,. \qquad [30.12]$$

Since

$$\frac{1}{2}\iint \frac{\varrho_P \varrho_Q}{r_{PQ}}\,\mathrm{d}V_P\,\mathrm{d}V_Q = E_{el}$$

represents the electrostatic energy (see Eq. [4.5]), μ can also be written as

$$\mu = \frac{4}{3}\frac{E_{el}}{c^2}. \qquad [30.13]$$

Because $\boldsymbol{K}^{(0)}$ has the character of an inertial force, *μ can be interpreted as an electromagnetic mass.* If the charge is considered to be localized within a sphere of radius a, then E_{el} is of the order of magnitude of e^2/a, where e is the electron charge (cf. Section 5).

For $\boldsymbol{K}^{(1)}$ there results

$$\boldsymbol{K}^{(1)} = +\frac{1}{c^3}\int \varrho_P \,\mathrm{d}V_P \cdot \frac{\partial^2}{\partial t^2}\int \varrho_Q \left\{\boldsymbol{v} - \frac{1}{6}\,\mathrm{grad}_P\,(\boldsymbol{v}\cdot\mathrm{grad}_P r_{PQ}^2)\right\}\mathrm{d}V_Q\,.$$

Since

$$\mathrm{grad}_P\,(\boldsymbol{v}\cdot\mathrm{grad}_P r_{PQ}^2) = 2\boldsymbol{v}$$

and

$$\int \varrho_P\,\mathrm{d}V_P = \int \varrho_Q \mathrm{d}V_Q = e\,,$$

therefore

$$\boldsymbol{K}^{(1)} = \frac{2}{3}\frac{e^2}{c^3}\ddot{\boldsymbol{v}} \qquad [30.14]$$

is the *radiation damping force.* It is *completely independent of the charge distribution.* (In particular, $a \to 0$ is permissible.)

The higher terms in the expansion of Eq. [30.9] are proportional to powers of a:

$$K^{(2)} \sim a\,, \qquad K^{(3)} \sim a^2\,.$$

These can be neglected if a is sufficiently small.

Thus, in addition to the damping force given by Eq. [30.14], still another term, that of Eq. [30.11], is obtained. However, if the equation of motion of the electron,

$$m\dot{\boldsymbol{v}} = \boldsymbol{K} = -\mu\dot{\boldsymbol{v}} + \frac{2}{3}\frac{e^2}{c^3}\ddot{\boldsymbol{v}} + \dots,$$

is written, then it is seen that m and μ, the mechanical and electromagnetic masses, respectively, cannot be separated, since this equation can also be written as

$$(m + \mu)\dot{\boldsymbol{v}} = \frac{2}{3}\frac{e^2}{c^3}\ddot{\boldsymbol{v}} + \dots. \qquad [30.15]$$

Hence, $m+\mu$ is simply the total mass of the electron.

For a point electron ($a=0$), the electromagnetic mass μ is infinite.

Bibliography

General Works

M. ABRAHAM and R. BECKER, *The Classical Theory of Electricity and Magnetism* (Hafner, New York, no date).

J. FRENKEL, *Lehrbuch der Elektrodynamik* (Berlin, 1926).

A. SOMMERFELD, *Lectures on Theoretical Physics, Vol. 3: Electrodynamics* (Academic Press, New York, 1952).

H. A. LORENTZ, *Vorlesungen über theoretische Physik* (Leipzig, 1907).

H. A. LORENTZ, Theorie der magneto-optischen Phänomene, *Encycl. math. Wiss.*, vol. 5, part 3 (Leipzig, 1909–1926).

J. C. MAXWELL, *A Treatise on Electricity and Magnetism* (Oxford, 1888; reprinted by Dover Publications, New York, 1954).

Special Works

H. A. LORENTZ, *Theory of Electrons and its Applications to the Phenomena of Light and Radiant Heat* (Leipzig, 1909; reprinted by Dover Publications, New York, 1952).

J. LARMOR, *Aether and Matter* (Cambridge, 1900).

Supplementary Bibliography

R. BECKER, *Electromagnetic Fields and Interactions*, edited by F. Sauter (Blaisdell, New York, 1964), 2 vols. (Supersedes M. ABRAHAM and R. BECKER, *The Classical Theory of Electricity and Magnetism*.)

J. D. JACKSON, *Classical Electrodynamics* (Wiley, New York, 1962).

L. D. LANDAU and E. M. LIFSHITZ, *Electrodynamics of Continuous Media* (Pergamon, New York, 1960).

W. K. H. PANOFSKY and M. PHILLIPS, *Classical Electricity and Magnetism* (Addison-Wesley, Reading, Mass., 1962).

W. R. SMYTHE, *Static and Dynamic Electricity* (McGraw-Hill, New York, 1950).

J. A. STRATTON, *Electromagnetic Theory* (McGraw-Hill, New York, 1941).

Appendix. Comments by the Editor

[A-1] (pp. 2, 3, 6, 123). These passages, although not written by Pauli himself, reflect Pauli's deepest convictions about the fundamental problems of theoretical physics, convictions that he repeatedly expressed during all his scientific life. This observation is substantiated by the following references to Pauli's works (volume and page numbers in square brackets refer to *Collected Scientific Papers by Wolfgang Pauli*, edited by R. Kronig and V. F. Weisskopf, John Wiley & Sons, Inc. (New York, 1964)):

Verhandl. Deutsche Physik. Ges. **21**, 742–750 (1919) [vol. 2, pp. 8–9].

Encyclopädie der mathematischen Wiss., vol. 5, Part. 2, B. G. Teubner (Leipzig, 1921), pp. 539–775 [vol. 1, p. 237].

Scientia **59**, 65–76 (1936) [vol. 2, pp. 747–748].

Prix Nobel 1946, pp. 131–147 (Stockholm, 1948) [vol. 2, p. 1095].

Helv. Phys. Acta, Suppl. 4, 261–267 (1956) [vol. 2, pp. 1304–1306].

Universitas **13**, 593–598 (1958) [vol. 2, p. 1367].

[A-2] (pp. 29, 37, 41, 55). The general electro- and magnetostatic property of matter is to be dielectric and diamagnetic, $\varepsilon > 1$ and $\mu < 1$. This is reflected by temperature independent ε and μ (except for the diamagnetism of semiconductors), that is, ε and μ are not of statistical origin. This case corresponds to the second analogy of p. 57; and $\varepsilon - 1 > 0$ and $(1/\mu) - 1 > 0$ express the fact that work has to be supplied to polarize the medium.

In the case of molecular or atomic (but nonpermanent) electric or magnetic moments, the first analogy of p. 57 applies; $\varepsilon - 1 > 0$ and $\mu - 1 > 0$ are due to alignment in the applied field and again express the fact that work has to be supplied to polarize the medium. Here ε and μ are of statistical origin and therefore temperature dependent.

In the case of a permanent electric or magnetic moment, $\boldsymbol{P}$ or $\boldsymbol{M}$ is a nonlinear function of $\boldsymbol{E}$ (ferroelectricity) or $\boldsymbol{H}$ (ferro-, ferri-, antiferromagnetism), respectively.

[A-3] (pp. 64, 67). The boundary-value problem for a steady-state conductor, $\partial\varphi/\partial n = 0$ on the lateral surface, $\varphi = \varphi_1 =$ constant and $\varphi = \varphi_2 =$ constant on the basal surfaces, has a unique solution φ. The proof is exactly the same as the one for the electrostatic problem (p. 41).

For cylindrical symmetry, φ is independent of azimuthal angle (also called φ on pp. 65, 67), and since $\sigma =$ constant one verifies by inspection that $\varphi = -\boldsymbol{E}\cdot\boldsymbol{z}$, $\boldsymbol{E} =$ constant, is a solution. Hence $\boldsymbol{i} =$ constant.

[A-4] (pp. 147-152). In an interesting recent contribution to the theory, Rohrlich remarked that the conventional calculation described here and elsewhere in the literature is not covariant and therefore [30.13] is not the correct expression for the electromagnetic mass (F. Rohrlich, *Classical Charged Particles*, Addison-Wesley Publishing Co. (Reading, Mass., 1965)). The correct procedure yields, according to Rohrlich (his Eq. (6-21)),

$$P^{\alpha}_{\text{self}} = \mu v^{\alpha},$$

where P^{α}_{self} is the part of the mechanical momentum 4-vector of the particle due to self-interaction, v^{α} the velocity 4-vector, and

$$\mu = \frac{E_{\text{el}}}{c^2}.$$

This is much more satisfactory than [30.13] since it satisfies the Einstein mass-energy relation.

Index

A CATALOG OF SELECTED

DOVER BOOKS

IN SCIENCE AND MATHEMATICS

Astronomy

BURNHAM'S CELESTIAL HANDBOOK, Robert Burnham, Jr. Thorough guide to the stars beyond our solar system. Exhaustive treatment. Alphabetical by constellation: Andromeda to Cetus in Vol. 1; Chamaeleon to Orion in Vol. 2; and Pavo to Vulpecula in Vol. 3. Hundreds of illustrations. Index in Vol. 3. 2,000pp. 6⅛ x 9¼.
Vol. I: 0-486-23567-X
Vol. II: 0-486-23568-8
Vol. III: 0-486-23673-0

EXPLORING THE MOON THROUGH BINOCULARS AND SMALL TELESCOPES, Ernest H. Cherrington, Jr. Informative, profusely illustrated guide to locating and identifying craters, rills, seas, mountains, other lunar features. Newly revised and updated with special section of new photos. Over 100 photos and diagrams. 240pp. 8¼ x 11. 0-486-24491-1

THE EXTRATERRESTRIAL LIFE DEBATE, 1750–1900, Michael J. Crowe. First detailed, scholarly study in English of the many ideas that developed from 1750 to 1900 regarding the existence of intelligent extraterrestrial life. Examines ideas of Kant, Herschel, Voltaire, Percival Lowell, many other scientists and thinkers. 16 illustrations. 704pp. 5⅜ x 8½. 0-486-40675-X

THEORIES OF THE WORLD FROM ANTIQUITY TO THE COPERNICAN REVOLUTION, Michael J. Crowe. Newly revised edition of an accessible, enlightening book recreates the change from an earth-centered to a sun-centered conception of the solar system. 242pp. 5⅜ x 8½. 0-486-41444-2

A HISTORY OF ASTRONOMY, A. Pannekoek. Well-balanced, carefully reasoned study covers such topics as Ptolemaic theory, work of Copernicus, Kepler, Newton, Eddington's work on stars, much more. Illustrated. References. 521pp. 5⅜ x 8½.
0-486-65994-1

A COMPLETE MANUAL OF AMATEUR ASTRONOMY: TOOLS AND TECHNIQUES FOR ASTRONOMICAL OBSERVATIONS, P. Clay Sherrod with Thomas L. Koed. Concise, highly readable book discusses: selecting, setting up and maintaining a telescope; amateur studies of the sun; lunar topography and occultations; observations of Mars, Jupiter, Saturn, the minor planets and the stars; an introduction to photoelectric photometry; more. 1981 ed. 124 figures. 25 halftones. 37 tables. 335pp. 6½ x 9¼. 0-486-40675-X

AMATEUR ASTRONOMER'S HANDBOOK, J. B. Sidgwick. Timeless, comprehensive coverage of telescopes, mirrors, lenses, mountings, telescope drives, micrometers, spectroscopes, more. 189 illustrations. 576pp. 5⅝ x 8¼. (Available in U.S. only.)
0-486-24034-7

STARS AND RELATIVITY, Ya. B. Zel'dovich and I. D. Novikov. Vol. 1 of *Relativistic Astrophysics* by famed Russian scientists. General relativity, properties of matter under astrophysical conditions, stars, and stellar systems. Deep physical insights, clear presentation. 1971 edition. References. 544pp. 5⅝ x 8¼. 0-486-69424-0

Chemistry

THE SCEPTICAL CHYMIST: THE CLASSIC 1661 TEXT, Robert Boyle. Boyle defines the term "element," asserting that all natural phenomena can be explained by the motion and organization of primary particles. 1911 ed. viii+232pp. 5⅜ x 8½.
0-486-42825-7

RADIOACTIVE SUBSTANCES, Marie Curie. Here is the celebrated scientist's doctoral thesis, the prelude to her receipt of the 1903 Nobel Prize. Curie discusses establishing atomic character of radioactivity found in compounds of uranium and thorium; extraction from pitchblende of polonium and radium; isolation of pure radium chloride; determination of atomic weight of radium; plus electric, photographic, luminous, heat, color effects of radioactivity. ii+94pp. 5⅜ x 8½. 0-486-42550-9

CHEMICAL MAGIC, Leonard A. Ford. Second Edition, Revised by E. Winston Grundmeier. Over 100 unusual stunts demonstrating cold fire, dust explosions, much more. Text explains scientific principles and stresses safety precautions. 128pp. 5⅜ x 8½. 0-486-67628-5

THE DEVELOPMENT OF MODERN CHEMISTRY, Aaron J. Ihde. Authoritative history of chemistry from ancient Greek theory to 20th-century innovation. Covers major chemists and their discoveries. 209 illustrations. 14 tables. Bibliographies. Indices. Appendices. 851pp. 5⅜ x 8½. 0-486-64235-6

CATALYSIS IN CHEMISTRY AND ENZYMOLOGY, William P. Jencks. Exceptionally clear coverage of mechanisms for catalysis, forces in aqueous solution, carbonyl- and acyl-group reactions, practical kinetics, more. 864pp. 5⅜ x 8½.
0-486-65460-5

ELEMENTS OF CHEMISTRY, Antoine Lavoisier. Monumental classic by founder of modern chemistry in remarkable reprint of rare 1790 Kerr translation. A must for every student of chemistry or the history of science. 539pp. 5⅜ x 8½. 0-486-64624-6

THE HISTORICAL BACKGROUND OF CHEMISTRY, Henry M. Leicester. Evolution of ideas, not individual biography. Concentrates on formulation of a coherent set of chemical laws. 260pp. 5⅜ x 8½. 0-486-61053-5

A SHORT HISTORY OF CHEMISTRY, J. R. Partington. Classic exposition explores origins of chemistry, alchemy, early medical chemistry, nature of atmosphere, theory of valency, laws and structure of atomic theory, much more. 428pp. 5⅜ x 8½. (Available in U.S. only.) 0-486-65977-1

GENERAL CHEMISTRY, Linus Pauling. Revised 3rd edition of classic first-year text by Nobel laureate. Atomic and molecular structure, quantum mechanics, statistical mechanics, thermodynamics correlated with descriptive chemistry. Problems. 992pp. 5⅜ x 8½. 0-486-65622-5

FROM ALCHEMY TO CHEMISTRY, John Read. Broad, humanistic treatment focuses on great figures of chemistry and ideas that revolutionized the science. 50 illustrations. 240pp. 5⅜ x 8½. 0-486-28690-8

Engineering

DE RE METALLICA, Georgius Agricola. The famous Hoover translation of greatest treatise on technological chemistry, engineering, geology, mining of early modern times (1556). All 289 original woodcuts. 638pp. 6¾ x 11. 0-486-60006-8

FUNDAMENTALS OF ASTRODYNAMICS, Roger Bate et al. Modern approach developed by U.S. Air Force Academy. Designed as a first course. Problems, exercises. Numerous illustrations. 455pp. 5⅜ x 8½. 0-486-60061-0

DYNAMICS OF FLUIDS IN POROUS MEDIA, Jacob Bear. For advanced students of ground water hydrology, soil mechanics and physics, drainage and irrigation engineering and more. 335 illustrations. Exercises, with answers. 784pp. 6⅛ x 9¼. 0-486-65675-6

THEORY OF VISCOELASTICITY (Second Edition), Richard M. Christensen. Complete consistent description of the linear theory of the viscoelastic behavior of materials. Problem-solving techniques discussed. 1982 edition. 29 figures. xiv+364pp. 6⅛ x 9¼. 0-486-42880-X

MECHANICS, J. P. Den Hartog. A classic introductory text or refresher. Hundreds of applications and design problems illuminate fundamentals of trusses, loaded beams and cables, etc. 334 answered problems. 462pp. 5⅜ x 8½. 0-486-60754-2

MECHANICAL VIBRATIONS, J. P. Den Hartog. Classic textbook offers lucid explanations and illustrative models, applying theories of vibrations to a variety of practical industrial engineering problems. Numerous figures. 233 problems, solutions. Appendix. Index. Preface. 436pp. 5⅜ x 8½. 0-486-64785-4

STRENGTH OF MATERIALS, J. P. Den Hartog. Full, clear treatment of basic material (tension, torsion, bending, etc.) plus advanced material on engineering methods, applications. 350 answered problems. 323pp. 5⅜ x 8½. 0-486-60755-0

A HISTORY OF MECHANICS, René Dugas. Monumental study of mechanical principles from antiquity to quantum mechanics. Contributions of ancient Greeks, Galileo, Leonardo, Kepler, Lagrange, many others. 671pp. 5⅜ x 8½. 0-486-65632-2

STABILITY THEORY AND ITS APPLICATIONS TO STRUCTURAL MECHANICS, Clive L. Dym. Self-contained text focuses on Koiter postbuckling analyses, with mathematical notions of stability of motion. Basing minimum energy principles for static stability upon dynamic concepts of stability of motion, it develops asymptotic buckling and postbuckling analyses from potential energy considerations, with applications to columns, plates, and arches. 1974 ed. 208pp. 5⅜ x 8½. 0-486-42541-X

METAL FATIGUE, N. E. Frost, K. J. Marsh, and L. P. Pook. Definitive, clearly written, and well-illustrated volume addresses all aspects of the subject, from the historical development of understanding metal fatigue to vital concepts of the cyclic stress that causes a crack to grow. Includes 7 appendixes. 544pp. 5⅜ x 8½. 0-486-40927-9

ROCKETS, Robert Goddard. Two of the most significant publications in the history of rocketry and jet propulsion: "A Method of Reaching Extreme Altitudes" (1919) and "Liquid Propellant Rocket Development" (1936). 128pp. 5⅜ x 8½. 0-486-42537-1

STATISTICAL MECHANICS: PRINCIPLES AND APPLICATIONS, Terrell L. Hill. Standard text covers fundamentals of statistical mechanics, applications to fluctuation theory, imperfect gases, distribution functions, more. 448pp. 5⅜ x 8½. 0-486-65390-0

ENGINEERING AND TECHNOLOGY 1650–1750: ILLUSTRATIONS AND TEXTS FROM ORIGINAL SOURCES, Martin Jensen. Highly readable text with more than 200 contemporary drawings and detailed engravings of engineering projects dealing with surveying, leveling, materials, hand tools, lifting equipment, transport and erection, piling, bailing, water supply, hydraulic engineering, and more. Among the specific projects outlined-transporting a 50-ton stone to the Louvre, erecting an obelisk, building timber locks, and dredging canals. 207pp. 8⅜ x 11¼. 0-486-42232-1

THE VARIATIONAL PRINCIPLES OF MECHANICS, Cornelius Lanczos. Graduate level coverage of calculus of variations, equations of motion, relativistic mechanics, more. First inexpensive paperbound edition of classic treatise. Index. Bibliography. 418pp. 5⅜ x 8½. 0-486-65067-7

PROTECTION OF ELECTRONIC CIRCUITS FROM OVERVOLTAGES, Ronald B. Standler. Five-part treatment presents practical rules and strategies for circuits designed to protect electronic systems from damage by transient overvoltages. 1989 ed. xxiv+434pp. 6⅛ x 9¼. 0-486-42552-5

ROTARY WING AERODYNAMICS, W. Z. Stepniewski. Clear, concise text covers aerodynamic phenomena of the rotor and offers guidelines for helicopter performance evaluation. Originally prepared for NASA. 537 figures. 640pp. 6⅛ x 9¼. 0-486-64647-5

INTRODUCTION TO SPACE DYNAMICS, William Tyrrell Thomson. Comprehensive, classic introduction to space-flight engineering for advanced undergraduate and graduate students. Includes vector algebra, kinematics, transformation of coordinates. Bibliography. Index. 352pp. 5⅜ x 8½. 0-486-65113-4

HISTORY OF STRENGTH OF MATERIALS, Stephen P. Timoshenko. Excellent historical survey of the strength of materials with many references to the theories of elasticity and structure. 245 figures. 452pp. 5⅜ x 8½. 0-486-61187-6

ANALYTICAL FRACTURE MECHANICS, David J. Unger. Self-contained text supplements standard fracture mechanics texts by focusing on analytical methods for determining crack-tip stress and strain fields. 336pp. 6⅛ x 9¼. 0-486-41737-9

STATISTICAL MECHANICS OF ELASTICITY, J. H. Weiner. Advanced, self-contained treatment illustrates general principles and elastic behavior of solids. Part 1, based on classical mechanics, studies thermoelastic behavior of crystalline and polymeric solids. Part 2, based on quantum mechanics, focuses on interatomic force laws, behavior of solids, and thermally activated processes. For students of physics and chemistry and for polymer physicists. 1983 ed. 96 figures. 496pp. 5⅜ x 8½. 0-486-42260-7

Mathematics

FUNCTIONAL ANALYSIS (Second Corrected Edition), George Bachman and Lawrence Narici. Excellent treatment of subject geared toward students with background in linear algebra, advanced calculus, physics and engineering. Text covers introduction to inner-product spaces, normed, metric spaces, and topological spaces; complete orthonormal sets, the Hahn-Banach Theorem and its consequences, and many other related subjects. 1966 ed. 544pp. 6⅛ x 9¼. 0-486-40251-7

ASYMPTOTIC EXPANSIONS OF INTEGRALS, Norman Bleistein & Richard A. Handelsman. Best introduction to important field with applications in a variety of scientific disciplines. New preface. Problems. Diagrams. Tables. Bibliography. Index. 448pp. 5⅜ x 8½. 0-486-65082-0

VECTOR AND TENSOR ANALYSIS WITH APPLICATIONS, A. I. Borisenko and I. E. Tarapov. Concise introduction. Worked-out problems, solutions, exercises. 257pp. 5⅝ x 8¼. 0-486-63833-2

AN INTRODUCTION TO ORDINARY DIFFERENTIAL EQUATIONS, Earl A. Coddington. A thorough and systematic first course in elementary differential equations for undergraduates in mathematics and science, with many exercises and problems (with answers). Index. 304pp. 5⅜ x 8½. 0-486-65942-9

FOURIER SERIES AND ORTHOGONAL FUNCTIONS, Harry F. Davis. An incisive text combining theory and practical example to introduce Fourier series, orthogonal functions and applications of the Fourier method to boundary-value problems. 570 exercises. Answers and notes. 416pp. 5⅜ x 8½. 0-486-65973-9

COMPUTABILITY AND UNSOLVABILITY, Martin Davis. Classic graduate-level introduction to theory of computability, usually referred to as theory of recurrent functions. New preface and appendix. 288pp. 5⅜ x 8½. 0-486-61471-9

ASYMPTOTIC METHODS IN ANALYSIS, N. G. de Bruijn. An inexpensive, comprehensive guide to asymptotic methods–the pioneering work that teaches by explaining worked examples in detail. Index. 224pp. 5⅜ x 8½ 0-486-64221-6

APPLIED COMPLEX VARIABLES, John W. Dettman. Step-by-step coverage of fundamentals of analytic function theory–plus lucid exposition of five important applications: Potential Theory; Ordinary Differential Equations; Fourier Transforms; Laplace Transforms; Asymptotic Expansions. 66 figures. Exercises at chapter ends. 512pp. 5⅜ x 8½. 0-486-64670-X

INTRODUCTION TO LINEAR ALGEBRA AND DIFFERENTIAL EQUATIONS, John W. Dettman. Excellent text covers complex numbers, determinants, orthonormal bases, Laplace transforms, much more. Exercises with solutions. Undergraduate level. 416pp. 5⅜ x 8½. 0-486-65191-6

RIEMANN'S ZETA FUNCTION, H. M. Edwards. Superb, high-level study of landmark 1859 publication entitled "On the Number of Primes Less Than a Given Magnitude" traces developments in mathematical theory that it inspired. xiv+315pp. 5⅜ x 8½. 0-486-41740-9

CALCULUS OF VARIATIONS WITH APPLICATIONS, George M. Ewing. Applications-oriented introduction to variational theory develops insight and promotes understanding of specialized books, research papers. Suitable for advanced undergraduate/graduate students as primary, supplementary text. 352pp. 5⅜ x 8½. 0-486-64856-7

COMPLEX VARIABLES, Francis J. Flanigan. Unusual approach, delaying complex algebra till harmonic functions have been analyzed from real variable viewpoint. Includes problems with answers. 364pp. 5⅜ x 8½. 0-486-61388-7

AN INTRODUCTION TO THE CALCULUS OF VARIATIONS, Charles Fox. Graduate-level text covers variations of an integral, isoperimetrical problems, least action, special relativity, approximations, more. References. 279pp. 5⅜ x 8½. 0-486-65499-0

COUNTEREXAMPLES IN ANALYSIS, Bernard R. Gelbaum and John M. H. Olmsted. These counterexamples deal mostly with the part of analysis known as "real variables." The first half covers the real number system, and the second half encompasses higher dimensions. 1962 edition. xxiv+198pp. 5⅜ x 8½. 0-486-42875-3

CATASTROPHE THEORY FOR SCIENTISTS AND ENGINEERS, Robert Gilmore. Advanced-level treatment describes mathematics of theory grounded in the work of Poincaré, R. Thom, other mathematicians. Also important applications to problems in mathematics, physics, chemistry and engineering. 1981 edition. References. 28 tables. 397 black-and-white illustrations. xvii + 666pp. 6⅛ x 9¼. 0-486-67539-4

INTRODUCTION TO DIFFERENCE EQUATIONS, Samuel Goldberg. Exceptionally clear exposition of important discipline with applications to sociology, psychology, economics. Many illustrative examples; over 250 problems. 260pp. 5⅜ x 8½. 0-486-65084-7

NUMERICAL METHODS FOR SCIENTISTS AND ENGINEERS, Richard Hamming. Classic text stresses frequency approach in coverage of algorithms, polynomial approximation, Fourier approximation, exponential approximation, other topics. Revised and enlarged 2nd edition. 721pp. 5⅜ x 8½. 0-486-65241-6

INTRODUCTION TO NUMERICAL ANALYSIS (2nd Edition), F. B. Hildebrand. Classic, fundamental treatment covers computation, approximation, interpolation, numerical differentiation and integration, other topics. 150 new problems. 669pp. 5⅜ x 8½. 0-486-65363-3

THREE PEARLS OF NUMBER THEORY, A. Y. Khinchin. Three compelling puzzles require proof of a basic law governing the world of numbers. Challenges concern van der Waerden's theorem, the Landau-Schnirelmann hypothesis and Mann's theorem, and a solution to Waring's problem. Solutions included. 64pp. 5⅜ x 8½. 0-486-40026-3

THE PHILOSOPHY OF MATHEMATICS: AN INTRODUCTORY ESSAY, Stephan Körner. Surveys the views of Plato, Aristotle, Leibniz & Kant concerning propositions and theories of applied and pure mathematics. Introduction. Two appendices. Index. 198pp. 5⅜ x 8½. 0-486-25048-2

INTRODUCTORY REAL ANALYSIS, A.N. Kolmogorov, S. V. Fomin. Translated by Richard A. Silverman. Self-contained, evenly paced introduction to real and functional analysis. Some 350 problems. 403pp. 5⅜ x 8½. 0-486-61226-0

APPLIED ANALYSIS, Cornelius Lanczos. Classic work on analysis and design of finite processes for approximating solution of analytical problems. Algebraic equations, matrices, harmonic analysis, quadrature methods, much more. 559pp. 5⅜ x 8½. 0-486-65656-X

AN INTRODUCTION TO ALGEBRAIC STRUCTURES, Joseph Landin. Superb self-contained text covers "abstract algebra": sets and numbers, theory of groups, theory of rings, much more. Numerous well-chosen examples, exercises. 247pp. 5⅜ x 8½. 0-486-65940-2

QUALITATIVE THEORY OF DIFFERENTIAL EQUATIONS, V. V. Nemytskii and V.V. Stepanov. Classic graduate-level text by two prominent Soviet mathematicians covers classical differential equations as well as topological dynamics and ergodic theory. Bibliographies. 523pp. 5⅜ x 8½. 0-486-65954-2

THEORY OF MATRICES, Sam Perlis. Outstanding text covering rank, nonsingularity and inverses in connection with the development of canonical matrices under the relation of equivalence, and without the intervention of determinants. Includes exercises. 237pp. 5⅜ x 8½. 0-486-66810-X

INTRODUCTION TO ANALYSIS, Maxwell Rosenlicht. Unusually clear, accessible coverage of set theory, real number system, metric spaces, continuous functions, Riemann integration, multiple integrals, more. Wide range of problems. Undergraduate level. Bibliography. 254pp. 5⅜ x 8½. 0-486-65038-3

MODERN NONLINEAR EQUATIONS, Thomas L. Saaty. Emphasizes practical solution of problems; covers seven types of equations. ". . . a welcome contribution to the existing literature...."–*Math Reviews*. 490pp. 5⅜ x 8½. 0-486-64232-1

MATRICES AND LINEAR ALGEBRA, Hans Schneider and George Phillip Barker. Basic textbook covers theory of matrices and its applications to systems of linear equations and related topics such as determinants, eigenvalues and differential equations. Numerous exercises. 432pp. 5⅜ x 8½. 0-486-66014-1

LINEAR ALGEBRA, Georgi E. Shilov. Determinants, linear spaces, matrix algebras, similar topics. For advanced undergraduates, graduates. Silverman translation. 387pp. 5⅜ x 8½. 0-486-63518-X

ELEMENTS OF REAL ANALYSIS, David A. Sprecher. Classic text covers fundamental concepts, real number system, point sets, functions of a real variable, Fourier series, much more. Over 500 exercises. 352pp. 5⅜ x 8½. 0-486-65385-4

SET THEORY AND LOGIC, Robert R. Stoll. Lucid introduction to unified theory of mathematical concepts. Set theory and logic seen as tools for conceptual understanding of real number system. 496pp. 5⅜ x 8¼. 0-486-63829-4

TENSOR CALCULUS, J.L. Synge and A. Schild. Widely used introductory text covers spaces and tensors, basic operations in Riemannian space, non-Riemannian spaces, etc. 324pp. 5⅜ x 8¼. 0-486-63612-7

ORDINARY DIFFERENTIAL EQUATIONS, Morris Tenenbaum and Harry Pollard. Exhaustive survey of ordinary differential equations for undergraduates in mathematics, engineering, science. Thorough analysis of theorems. Diagrams. Bibliography. Index. 818pp. 5⅜ x 8½. 0-486-64940-7

INTEGRAL EQUATIONS, F. G. Tricomi. Authoritative, well-written treatment of extremely useful mathematical tool with wide applications. Volterra Equations, Fredholm Equations, much more. Advanced undergraduate to graduate level. Exercises. Bibliography. 238pp. 5⅜ x 8½. 0-486-64828-1

FOURIER SERIES, Georgi P. Tolstov. Translated by Richard A. Silverman. A valuable addition to the literature on the subject, moving clearly from subject to subject and theorem to theorem. 107 problems, answers. 336pp. 5⅜ x 8½. 0-486-63317-9

INTRODUCTION TO MATHEMATICAL THINKING, Friedrich Waismann. Examinations of arithmetic, geometry, and theory of integers; rational and natural numbers; complete induction; limit and point of accumulation; remarkable curves; complex and hypercomplex numbers, more. 1959 ed. 27 figures. xii+260pp. 5⅜ x 8½. 0-486-63317-9

POPULAR LECTURES ON MATHEMATICAL LOGIC, Hao Wang. Noted logician's lucid treatment of historical developments, set theory, model theory, recursion theory and constructivism, proof theory, more. 3 appendixes. Bibliography. 1981 edition. ix + 283pp. 5⅜ x 8½. 0-486-67632-3

CALCULUS OF VARIATIONS, Robert Weinstock. Basic introduction covering isoperimetric problems, theory of elasticity, quantum mechanics, electrostatics, etc. Exercises throughout. 326pp. 5⅜ x 8½. 0-486-63069-2

THE CONTINUUM: A CRITICAL EXAMINATION OF THE FOUNDATION OF ANALYSIS, Hermann Weyl. Classic of 20th-century foundational research deals with the conceptual problem posed by the continuum. 156pp. 5⅜ x 8½. 0-486-67982-9

CHALLENGING MATHEMATICAL PROBLEMS WITH ELEMENTARY SOLUTIONS, A. M. Yaglom and I. M. Yaglom. Over 170 challenging problems on probability theory, combinatorial analysis, points and lines, topology, convex polygons, many other topics. Solutions. Total of 445pp. 5⅜ x 8½. Two-vol. set. Vol. I: 0-486-65536-9 Vol. II: 0-486-65537-7

INTRODUCTION TO PARTIAL DIFFERENTIAL EQUATIONS WITH APPLICATIONS, E. C. Zachmanoglou and Dale W. Thoe. Essentials of partial differential equations applied to common problems in engineering and the physical sciences. Problems and answers. 416pp. 5⅜ x 8½. 0-486-65251-3

THE THEORY OF GROUPS, Hans J. Zassenhaus. Well-written graduate-level text acquaints reader with group-theoretic methods and demonstrates their usefulness in mathematics. Axioms, the calculus of complexes, homomorphic mapping, *p*-group theory, more. 276pp. 5⅜ x 8½. 0-486-40922-8

Math–Decision Theory, Statistics, Probability

ELEMENTARY DECISION THEORY, Herman Chernoff and Lincoln E. Moses. Clear introduction to statistics and statistical theory covers data processing, probability and random variables, testing hypotheses, much more. Exercises. 364pp. 5⅜ x 8½. 0-486-65218-1

STATISTICS MANUAL, Edwin L. Crow et al. Comprehensive, practical collection of classical and modern methods prepared by U.S. Naval Ordnance Test Station. Stress on use. Basics of statistics assumed. 288pp. 5⅜ x 8½. 0-486-60599-X

SOME THEORY OF SAMPLING, William Edwards Deming. Analysis of the problems, theory and design of sampling techniques for social scientists, industrial managers and others who find statistics important at work. 61 tables. 90 figures. xvii +602pp. 5⅜ x 8½. 0-486-64684-X

LINEAR PROGRAMMING AND ECONOMIC ANALYSIS, Robert Dorfman, Paul A. Samuelson and Robert M. Solow. First comprehensive treatment of linear programming in standard economic analysis. Game theory, modern welfare economics, Leontief input-output, more. 525pp. 5⅜ x 8½. 0-486-65491-5

PROBABILITY: AN INTRODUCTION, Samuel Goldberg. Excellent basic text covers set theory, probability theory for finite sample spaces, binomial theorem, much more. 360 problems. Bibliographies. 322pp. 5⅜ x 8½. 0-486-65252-1

GAMES AND DECISIONS: INTRODUCTION AND CRITICAL SURVEY, R. Duncan Luce and Howard Raiffa. Superb nontechnical introduction to game theory, primarily applied to social sciences. Utility theory, zero-sum games, n-person games, decision-making, much more. Bibliography. 509pp. 5⅜ x 8½. 0-486-65943-7

INTRODUCTION TO THE THEORY OF GAMES, J. C. C. McKinsey. This comprehensive overview of the mathematical theory of games illustrates applications to situations involving conflicts of interest, including economic, social, political, and military contexts. Appropriate for advanced undergraduate and graduate courses; advanced calculus a prerequisite. 1952 ed. x+372pp. 5⅜ x 8½. 0-486-42811-7

FIFTY CHALLENGING PROBLEMS IN PROBABILITY WITH SOLUTIONS, Frederick Mosteller. Remarkable puzzlers, graded in difficulty, illustrate elementary and advanced aspects of probability. Detailed solutions. 88pp. 5⅜ x 8½. 65355-2

PROBABILITY THEORY: A CONCISE COURSE, Y. A. Rozanov. Highly readable, self-contained introduction covers combination of events, dependent events, Bernoulli trials, etc. 148pp. 5⅜ x 8¼. 0-486-63544-9

STATISTICAL METHOD FROM THE VIEWPOINT OF QUALITY CONTROL, Walter A. Shewhart. Important text explains regulation of variables, uses of statistical control to achieve quality control in industry, agriculture, other areas. 192pp. 5⅜ x 8½. 0-486-65232-7

Math–Geometry and Topology

ELEMENTARY CONCEPTS OF TOPOLOGY, Paul Alexandroff. Elegant, intuitive approach to topology from set-theoretic topology to Betti groups; how concepts of topology are useful in math and physics. 25 figures. 57pp. 5⅜ x 8½. 0-486-60747-X

COMBINATORIAL TOPOLOGY, P. S. Alexandrov. Clearly written, well-organized, three-part text begins by dealing with certain classic problems without using the formal techniques of homology theory and advances to the central concept, the Betti groups. Numerous detailed examples. 654pp. 5⅜ x 8½. 0-486-40179-0

EXPERIMENTS IN TOPOLOGY, Stephen Barr. Classic, lively explanation of one of the byways of mathematics. Klein bottles, Moebius strips, projective planes, map coloring, problem of the Koenigsberg bridges, much more, described with clarity and wit. 43 figures. 210pp. 5⅜ x 8½. 0-486-25933-1

THE GEOMETRY OF RENÉ DESCARTES, René Descartes. The great work founded analytical geometry. Original French text, Descartes's own diagrams, together with definitive Smith-Latham translation. 244pp. 5⅜ x 8½. 0-486-60068-8

EUCLIDEAN GEOMETRY AND TRANSFORMATIONS, Clayton W. Dodge. This introduction to Euclidean geometry emphasizes transformations, particularly isometries and similarities. Suitable for undergraduate courses, it includes numerous examples, many with detailed answers. 1972 ed. viii+296pp. 6⅛ x 9¼. 0-486-43476-1

PRACTICAL CONIC SECTIONS: THE GEOMETRIC PROPERTIES OF ELLIPSES, PARABOLAS AND HYPERBOLAS, J. W. Downs. This text shows how to create ellipses, parabolas, and hyperbolas. It also presents historical background on their ancient origins and describes the reflective properties and roles of curves in design applications. 1993 ed. 98 figures. xii+100pp. 6½ x 9¼. 0-486-42876-1

THE THIRTEEN BOOKS OF EUCLID'S ELEMENTS, translated with introduction and commentary by Sir Thomas L. Heath. Definitive edition. Textual and linguistic notes, mathematical analysis. 2,500 years of critical commentary. Unabridged. 1,414pp. 5⅜ x 8½.. Three-vol. set.
Vol. I: 0-486-60088-2 Vol. II: 0-486-60089-0 Vol. III: 0-486-60090-4

SPACE AND GEOMETRY: IN THE LIGHT OF PHYSIOLOGICAL, PSYCHOLOGICAL AND PHYSICAL INQUIRY, Ernst Mach. Three essays by an eminent philosopher and scientist explore the nature, origin, and development of our concepts of space, with a distinctness and precision suitable for undergraduate students and other readers. 1906 ed. vi+148pp. 5⅜ x 8½. 0-486-43909-7

GEOMETRY OF COMPLEX NUMBERS, Hans Schwerdtfeger. Illuminating, widely praised book on analytic geometry of circles, the Moebius transformation, and two-dimensional non-Euclidean geometries. 200pp. 5⅝ x 8¼. 0-486-63830-8

DIFFERENTIAL GEOMETRY, Heinrich W. Guggenheimer. Local differential geometry as an application of advanced calculus and linear algebra. Curvature, transformation groups, surfaces, more. Exercises. 62 figures. 378pp. 5⅜ x 8½. 0-486-63433-7

History of Math

THE WORKS OF ARCHIMEDES, Archimedes (T. L. Heath, ed.). Topics include the famous problems of the ratio of the areas of a cylinder and an inscribed sphere; the measurement of a circle; the properties of conoids, spheroids, and spirals; and the quadrature of the parabola. Informative introduction. clxxxvi+326pp. 5⅜ x 8½. 0-486-42084-1

A SHORT ACCOUNT OF THE HISTORY OF MATHEMATICS, W. W. Rouse Ball. One of clearest, most authoritative surveys from the Egyptians and Phoenicians through 19th-century figures such as Grassman, Galois, Riemann. Fourth edition. 522pp. 5⅜ x 8½. 0-486-20630-0

THE HISTORY OF THE CALCULUS AND ITS CONCEPTUAL DEVELOPMENT, Carl B. Boyer. Origins in antiquity, medieval contributions, work of Newton, Leibniz, rigorous formulation. Treatment is verbal. 346pp. 5⅜ x 8½. 0-486-60509-4

THE HISTORICAL ROOTS OF ELEMENTARY MATHEMATICS, Lucas N. H. Bunt, Phillip S. Jones, and Jack D. Bedient. Fundamental underpinnings of modern arithmetic, algebra, geometry and number systems derived from ancient civilizations. 320pp. 5⅜ x 8½. 0-486-25563-8

A HISTORY OF MATHEMATICAL NOTATIONS, Florian Cajori. This classic study notes the first appearance of a mathematical symbol and its origin, the competition it encountered, its spread among writers in different countries, its rise to popularity, its eventual decline or ultimate survival. Original 1929 two-volume edition presented here in one volume. xxviii+820pp. 5⅜ x 8½. 0-486-67766-4

GAMES, GODS & GAMBLING: A HISTORY OF PROBABILITY AND STATISTICAL IDEAS, F. N. David. Episodes from the lives of Galileo, Fermat, Pascal, and others illustrate this fascinating account of the roots of mathematics. Features thought-provoking references to classics, archaeology, biography, poetry. 1962 edition. 304pp. 5⅜ x 8½. (Available in U.S. only.) 0-486-40023-9

OF MEN AND NUMBERS: THE STORY OF THE GREAT MATHEMATICIANS, Jane Muir. Fascinating accounts of the lives and accomplishments of history's greatest mathematical minds–Pythagoras, Descartes, Euler, Pascal, Cantor, many more. Anecdotal, illuminating. 30 diagrams. Bibliography. 256pp. 5⅜ x 8½. 0-486-28973-7

HISTORY OF MATHEMATICS, David E. Smith. Nontechnical survey from ancient Greece and Orient to late 19th century; evolution of arithmetic, geometry, trigonometry, calculating devices, algebra, the calculus. 362 illustrations. 1,355pp. 5⅜ x 8½. Two-vol. set. Vol. I: 0-486-20429-4 Vol. II: 0-486-20430-8

A CONCISE HISTORY OF MATHEMATICS, Dirk J. Struik. The best brief history of mathematics. Stresses origins and covers every major figure from ancient Near East to 19th century. 41 illustrations. 195pp. 5⅜ x 8½. 0-486-60255-9

Physics

OPTICAL RESONANCE AND TWO-LEVEL ATOMS, L. Allen and J. H. Eberly. Clear, comprehensive introduction to basic principles behind all quantum optical resonance phenomena. 53 illustrations. Preface. Index. 256pp. 5⅜ x 8½. 0-486-65533-4

QUANTUM THEORY, David Bohm. This advanced undergraduate-level text presents the quantum theory in terms of qualitative and imaginative concepts, followed by specific applications worked out in mathematical detail. Preface. Index. 655pp. 5⅜ x 8½. 0-486-65969-0

ATOMIC PHYSICS (8th EDITION), Max Born. Nobel laureate's lucid treatment of kinetic theory of gases, elementary particles, nuclear atom, wave-corpuscles, atomic structure and spectral lines, much more. Over 40 appendices, bibliography. 495pp. 5⅜ x 8½. 0-486-65984-4

A SOPHISTICATE'S PRIMER OF RELATIVITY, P. W. Bridgman. Geared toward readers already acquainted with special relativity, this book transcends the view of theory as a working tool to answer natural questions: What is a frame of reference? What is a "law of nature"? What is the role of the "observer"? Extensive treatment, written in terms accessible to those without a scientific background. 1983 ed. xlviii+172pp. 5⅜ x 8½. 0-486-42549-5

AN INTRODUCTION TO HAMILTONIAN OPTICS, H. A. Buchdahl. Detailed account of the Hamiltonian treatment of aberration theory in geometrical optics. Many classes of optical systems defined in terms of the symmetries they possess. Problems with detailed solutions. 1970 edition. xv + 360pp. 5⅜ x 8½. 0-486-67597-1

PRIMER OF QUANTUM MECHANICS, Marvin Chester. Introductory text examines the classical quantum bead on a track: its state and representations; operator eigenvalues; harmonic oscillator and bound bead in a symmetric force field; and bead in a spherical shell. Other topics include spin, matrices, and the structure of quantum mechanics; the simplest atom; indistinguishable particles; and stationary-state perturbation theory. 1992 ed. xiv+314pp. 6⅛ x 9¼. 0-486-42878-8

LECTURES ON QUANTUM MECHANICS, Paul A. M. Dirac. Four concise, brilliant lectures on mathematical methods in quantum mechanics from Nobel Prize-winning quantum pioneer build on idea of visualizing quantum theory through the use of classical mechanics. 96pp. 5⅜ x 8½. 0-486-41713-1

THIRTY YEARS THAT SHOOK PHYSICS: THE STORY OF QUANTUM THEORY, George Gamow. Lucid, accessible introduction to influential theory of energy and matter. Careful explanations of Dirac's anti-particles, Bohr's model of the atom, much more. 12 plates. Numerous drawings. 240pp. 5⅜ x 8½. 0-486-24895-X

ELECTRONIC STRUCTURE AND THE PROPERTIES OF SOLIDS: THE PHYSICS OF THE CHEMICAL BOND, Walter A. Harrison. Innovative text offers basic understanding of the electronic structure of covalent and ionic solids, simple metals, transition metals and their compounds. Problems. 1980 edition. 582pp. 6⅛ x 9¼. 0-486-66021-4

HYDRODYNAMIC AND HYDROMAGNETIC STABILITY, S. Chandrasekhar. Lucid examination of the Rayleigh-Benard problem; clear coverage of the theory of instabilities causing convection. 704pp. 5⅜ x 8¼. 0-486-64071-X

INVESTIGATIONS ON THE THEORY OF THE BROWNIAN MOVEMENT, Albert Einstein. Five papers (1905–8) investigating dynamics of Brownian motion and evolving elementary theory. Notes by R. Fürth. 122pp. 5⅜ x 8½. 0-486-60304-0

THE PHYSICS OF WAVES, William C. Elmore and Mark A. Heald. Unique overview of classical wave theory. Acoustics, optics, electromagnetic radiation, more. Ideal as classroom text or for self-study. Problems. 477pp. 5⅜ x 8½. 0-486-64926-1

GRAVITY, George Gamow. Distinguished physicist and teacher takes reader-friendly look at three scientists whose work unlocked many of the mysteries behind the laws of physics: Galileo, Newton, and Einstein. Most of the book focuses on Newton's ideas, with a concluding chapter on post-Einsteinian speculations concerning the relationship between gravity and other physical phenomena. 160pp. 5⅜ x 8½. 0-486-42563-0

PHYSICAL PRINCIPLES OF THE QUANTUM THEORY, Werner Heisenberg. Nobel Laureate discusses quantum theory, uncertainty, wave mechanics, work of Dirac, Schroedinger, Compton, Wilson, Einstein, etc. 184pp. 5⅜ x 8½. 0-486-60113-7

ATOMIC SPECTRA AND ATOMIC STRUCTURE, Gerhard Herzberg. One of best introductions; especially for specialist in other fields. Treatment is physical rather than mathematical. 80 illustrations. 257pp. 5⅜ x 8½. 0-486-60115-3

AN INTRODUCTION TO STATISTICAL THERMODYNAMICS, Terrell L. Hill. Excellent basic text offers wide-ranging coverage of quantum statistical mechanics, systems of interacting molecules, quantum statistics, more. 523pp. 5⅜ x 8½. 0-486-65242-4

THEORETICAL PHYSICS, Georg Joos, with Ira M. Freeman. Classic overview covers essential math, mechanics, electromagnetic theory, thermodynamics, quantum mechanics, nuclear physics, other topics. First paperback edition. xxiii + 885pp. 5⅜ x 8½. 0-486-65227-0

PROBLEMS AND SOLUTIONS IN QUANTUM CHEMISTRY AND PHYSICS, Charles S. Johnson, Jr. and Lee G. Pedersen. Unusually varied problems, detailed solutions in coverage of quantum mechanics, wave mechanics, angular momentum, molecular spectroscopy, more. 280 problems plus 139 supplementary exercises. 430pp. 6½ x 9¼. 0-486-65236-X

THEORETICAL SOLID STATE PHYSICS, Vol. 1: Perfect Lattices in Equilibrium; Vol. II: Non-Equilibrium and Disorder, William Jones and Norman H. March. Monumental reference work covers fundamental theory of equilibrium properties of perfect crystalline solids, non-equilibrium properties, defects and disordered systems. Appendices. Problems. Preface. Diagrams. Index. Bibliography. Total of 1,301pp. 5⅜ x 8½. Two volumes. Vol. I: 0-486-65015-4 Vol. II: 0-486-65016-2

WHAT IS RELATIVITY? L. D. Landau and G. B. Rumer. Written by a Nobel Prize physicist and his distinguished colleague, this compelling book explains the special theory of relativity to readers with no scientific background, using such familiar objects as trains, rulers, and clocks. 1960 ed. vi+72pp. 5⅜ x 8½. 0-486-42806-0